T0136004

Isaac Physics Skills

Developing mastery of

Essential pre-university physics

A.C. Machacek
Assistant Head, Westcliff High School for Boys

J.J. Crowter
Head of Physics, Royal Grammar School, High Wycombe

Periphyseos Press
Cambridge, UK.

CAMBRIDGE
UNIVERSITY PRESS

Co-published in Cambridge, United Kingdom, by
Periphyseos Press and Cambridge University Press.

www.periphyseos.org.uk and www.cambridge.org

First published & reprinted 2015
Second edition & First reprint 2015
Second – Sixth reprints 2016, Seventh – Ninth reprints 2017, Tenth &
Eleventh reprints 2018
Third edition 2018. First reprint 2019
Co-published adaptation, 2020

Printed and bound in the UK by Short Run Press Limited, Exeter.

Typeset in LaTeX

A catalogue record for this publication is available from the British Library

ISBN 978-1-8382160-0-9 Paperback

Use this book in parallel with the electronic version at isaacphysics.org.
Marking of answers and compilation of results is free on Isaac Physics.
Register as a student or as a teacher to gain full functionality and support.

used with kind permission of M. J. Rutter.

Attaining Mastery – Notes for the Student and the Teacher

These sheets are of two kinds - skill sheets and fact sheets and are equally valuable in both the early stages of learning and in revision.

The skill sheets provide practice for the student in applying a single principle of physics to a range of reasonably straightforward situations - often starting out with the substitution of values into an equation. After the first few questions have enabled the student to gain confidence in using the equation, subsequent questions may require the use of more than one equation or principle, or insight to solve a novel problem.

The fact sheets test knowledge on the parts of sixth form physics courses which require reading. While eminently suitable for revision, they may also be used to verify that the student has performed prior reading on the operation of an MRI scanner, for example, before it is discussed in class.

While the manner of use is up to the teacher and the student, we recommend that until a pass mark (we suggest 75%, as indicated in the square by each skill sheet) is obtained, the student studies further, then repeats a selection of questions. This process is repeated until the student passes. This is our understanding of applying a mastery method to ensure a good foundation is laid for a pre-university physics education. The teacher's mark book records how many attempts the student has taken rather than the mark obtained on a sole attempt. Students, likewise, have a list of the skills, and tick them off when the required level of proficiency has been obtained. A grid is provided at the front of this book for this very purpose. In this way, all students achieve mastery of the skills, and do not move on until this has been achieved. We have found that all students who have the capacity to pass an A-level course have the capacity to attain mastery of all of these skills. However, we find that if such mastery has been obtained, the student goes on to gain a greater understanding of physics and a higher grade at A-level than would be expected otherwise.

$^n/_m$

We also recommend that an answer is not considered correct unless it is numerically accurate, is given to an appropriate number of significant figures, and incorporates a suitable unit: a student cannot be satisfied with their comprehension while they are still getting the final result incorrect. After all, this would be intolerable in any practical situation in which physics principles were applied. The Isaac Physics on-line version of this book, and its associated marking facility, follow these significant figure and unit requirements.

A mapping of each set of exercises in the book onto the 6 major school exam specifications in the UK is given on-line at isaacphysics.org/pages/syllabus_map where, with a click, the appropriate part of the book can be accessed and the problems solved or set as homework.

ACM & JJC
Westcliff-on-Sea & High Wycombe, 2015

Acknowledgements

These sheets, and the approach presented here for using them, were first devised for use in the Physics Department at the Royal Grammar School in High Wycombe, where we served as colleagues from 2010 to 2014. We are grateful to the students who have put them to such productive use, and have given valuable feedback. We are also grateful to colleagues in other schools who have also applied these methods and given us encouragement – particularly Keith Dalby at Westcliff High School for Boys.

However, our greatest thanks are to our colleagues in High Wycombe who tirelessly used these methods with their students to bring about better learning - to Peter Glendining, to Matthew Hale (who also devised an excellent spreadsheet for the monitoring of student progress on these sheets), and to Paula Dove who suggested the need for fact sheets.

We are also grateful to Prof. Mark Warner and his team at the Cavendish Laboratory for their assistance in turning a set of A4 problem sheets into a professional publication. Much credit is due to James Sharkey and Aleksandr Bowkis for the typesetting and editing. We are also grateful to Robin Hughes and Ally Davies for their thoughtful comments on the draft of this text.

Anton has three personal 'thank yous' to record: firstly to the Civil Aviation Authority. Its insistence that all student pilots gain mastery in all ground school topics inspired the development of a similar approach to A-level Physics teaching. Secondly, to Jennifer for her support and feedback in the implementation of the project, not to mention the enthusiasm with which she has championed its adoption, and her extensive written contributions. Most importantly, he wishes to acknowledge Helen Machacek. For not only has she remained the most supportive, encouraging and loving wife imaginable, but she bore much more than her fair share of the responsibility of looking after their children during the holiday periods when her husband was writing many of the sheets.

Jennifer wishes to thank Anton, for being such an inspiration, both as a physicist and as a friend. He has opened her eyes to so many ideas and possibilities in the teaching of Physics and in interactions with students, and this skills sheet approach is just one example of that. He has been unendingly supportive and his wisdom, advice and example have contributed hugely to the teacher and to the person she is today.

And finally, we thank you for being willing to try them out too. We wish you well with your studies, and hope that these sheets help you and your students attain understanding and success.

Soli Deo Gloria,

ACM & JJC
Westcliff-on-Sea & High Wycombe, 2015

Using Isaac Physics with this book

Isaac Physics offers on-line versions of each sheet at isaacphysics.org/books/
physics_skills_19 where a student can enter answers. This on-line tool will mark
answers, giving immediate feedback to a student who, if registered on isaacphys-
ics.org, can have their progress stored and even retrieved for their CV! Teachers can
set a sheet for class homework as the appropriate theme is being taught, and again
for pre-exam revision. Isaac Physics can return the fully assembled and analysed
marks to the teacher, if registered for this free service. Isaac Physics zealously follows
the significant figures (sf) rules below and warns if your answer has a sf problem.

Uncertainty and Significant Figures

In physics, numbers represent values that have uncertainty and this is indicated by
the number of significant figures in an answer.

Significant figures

When there is a decimal point (dp), all digits are significant, except leading (leftmost)
zeros: 2.00 (3 sf); 0.020 (2 sf); 200.1 (4 sf); 200.010 (6sf)

Numbers without a dp can have an *absolute accuracy*: 4 people; 3 electrons.

Some numbers can be ambiguous: 200 could be 1, 2 or 3 sf (see below). Assume
such numbers have the same number of s.f. as other numbers in the question.

Combining quantities

Multiplying or dividing numbers gives a result with a number of sf equal to that of
the number with the smallest number of sf:

$x = 2.31$, $y = 4.921$ gives $xy = 11.4$ (3 sf, the same as x).

An absolutely accurate number multiplied in does not influence the above.

Standard form

On-line, and sometimes in texts, one uses a letter 'x' in place of a times sign and ˆ
denotes "to the power of":

1800000 could be 1.80x10ˆ6 (3 sf) and 0.0000155 is 1.55x10ˆ-5

(standardly, 1.80×10^6 and 1.55×10^{-5})

The letter 'e' can denote "times 10 to the power of": 1.80e6 and 1.55e-5.

Significant figures in standard form

Standard form eliminates ambiguity: In $n.nnn \times 10^n$, the numbers before and after
the decimal point are significant:

$191 = 1.91 \times 10^2$ (3 sf); 191 is $190 = 1.9 \times 10^2$ (2 sf); 191 is $200 = 2 \times 10^2$ (1 sf).

Answers to questions

In this book and on-line, give the appropriate number of sf: For example, when the
least accurate data in a question is given to 3 significant figures, then the answer
should be given to three significant figures; see above. Too many sf are meaningless;
giving too few discards information. Exam boards also require consistency in sf.

Physical Quantities

Quantity	Magnitude	Unit
Permittivity of free space (ϵ_0)	8.85×10^{-12}	F m^{-1}
Electrostatic force constant ($1/4\pi\epsilon_0$)	8.99×10^9	$\text{N m}^2\ \text{C}^{-2}$
Speed of light (*in vacuo*)	3.00×10^8	m s^{-1}
Universal gravitational constant (G)	6.67×10^{-11}	$\text{N m}^2\ \text{kg}^{-2}$
Avogadro constant (N_A)	6.02×10^{23}	mol^{-1}
Boltzmann constant (k_B)	1.38×10^{-23}	J K^{-1}
Gas constant (R)	8.31	$\text{J mol}^{-1}\ \text{K}^{-1}$
Planck constant (h)	6.63×10^{-34}	J s
Charge of proton (e)	1.60×10^{-19}	C
Electron volt (eV)	1.60×10^{-19}	J
Unified atomic mass unit (u)	1.66054×10^{-27}	kg
Mass of neutron (m_n)	1.67493×10^{-27}	kg
Mass of neutron (m_n)	1.00867	u
Mass of proton (m_p)	1.67262×10^{-27}	kg
Mass of proton (m_p)	1.00728	u
Mass of electron (m_e)	9.10938×10^{-31}	kg
Mass of electron (m_e)	5.48580×10^{-4}	u
0 °C	273.15	K
1 parsec (pc)	3.086×10^{16}	m
Acceleration due to gravity (g)	9.81	m s^{-2}
Seconds per year	3.156×10^7	-
Light year	9.46×10^{15}	m
Specific heat capacity of water	4180	$\text{J kg}^{-1}\ \text{K}^{-1}$

Contents

Checklists

Unit	Skill	
A1	Choose an appropriate equation from a list, re-arrange it, substitute numbers for variables, and calculate the unknown quantity.	
A2	Express units in terms of SI base units.	
A3	Express a measurement in standard form (2.43×10^{-8}) or using a prefix (24.3 ns) to a given number of significant figures.	
A4	Convert measurements from one unit to another.	
A5	Calculate the gradient or y-intercept of a straight line on a graph and give its unit.	
A7	Estimate the area under the line on a graph and give its unit.	
A8	Estimate areas under lines using units with prefixes.	
A9	Calculate changes in response to changes by factors & percentages.	
A10	Calculate changes when a proportionality is known.	
B1	Determine the horizontal and vertical components of a vector (displacement, velocity, force or the electric field in Malus' Law).	
B2	Determine the sum of two vectors by scale drawing or trigonometry where the triangle of vectors always has a 90° angle.	
B3	Solve problems of uniform accelerated motion in 1-dimension (SUVAT problems).	
B4	Solve trajectory problems using the independence of horizontal and vertical motion.	
B5	Find the missing force in a question using the principle of moments.	
B6	Stress, strain and Young's modulus.	
B7	Force & energy calculations for springs separately & combined.	
B8	Mechanical calculations of energy and power.	
B9	Energy and force in springs and stretched materials.	
C1	Resistors & resistivity.	
C2	Understands the relationship between charge carrier motion and electric current.	
C4	Solve circuit problems using Kirchhoff's Laws.	
C5	Find the voltage across components in a potential divider circuit.	
C6	Work out the terminal p.d. of a battery given the current supplied, e.m.f. and internal resistance.	
D1	Calculate amplitudes and intensities from power.	
D2	Intensities of light after passing a polarizer.	
D3	Determine whether a wave, traveling by two different routes to a detector will interfere constructively or destructively.	

Unit	Skill	
D4	Select the right equation to use for two-source interference problems, re-arrange it, and solve it to obtain the correct answer.	
D5	Understand standing waves.	
D6	Select an equation to use in solving photoelectric effect problems, re-arrange it, and solve it to obtain the correct answer.	
D7	Perform calculations relevant to quantum physics.	
D8	Calculate refractive indices, angles of refraction, and critical angles.	
D9	Calculate electromagnetic spectra from atomic energy schemes.	
E1	Estimate absolute uncertainties.	
E2	Calculate relative uncertainties.	
E3	Estimate the relative uncertainty in a calculated result from the uncertainties of the original measurements.	
E4	Assess whether measurements are accurate or reliable.	
F1	Calculate the force needed to change an object's momentum in a given time.	
F2	Solve a 1-d problem in conservation of momentum.	
F3	Convert angles from degrees to radians, can convert ordinary speeds into angular speeds, can convert between frequency, time period & angular frequency.	
F4	Work out the size and direction of the force needed to keep an object in uniform circular motion.	
F5	Work out the gravitational force on an object using Newton's law of gravity, either directly (knowing M) or by comparison with another object where the force is known.	
F6	Work out the time period of a circular orbit from its radius (or vice versa) without looking up Kepler's 3rd Law.	
F7	Perform calculations related to oscillators.	
G1	Convert Celsius into kelvin, and know when K must be used.	
G2	Use the right form of the gas law ($pV=nRT$ or $pV = NkT$ or pV/T=const) to solve a problem involving gases.	
G3	Calculate energies required to cause temperature changes, and to calculate the final temperature of mixtures.	
G4	Calculate energies required to cause changes of state.	
H1	Work out the force (direction & magnitude) on an electron or alpha particle between two charged plates.	
H2	Calculate the electric field E near one or two point charges.	
H3	Find the speed of electrons accelerated from rest in an electric field.	

Unit	Skill	
H4	Work out the force (direction & magnitude) on a wire carrying a current in a magnetic field.	
H5	Work out the force (direction & magnitude) on a moving charged particle in a magnetic field.	
H6	Work out the radius of the circular path followed by a charged particle in a magnetic field.	
H7	Work out the e.m.f. (magnitude & direction) induced in a coil of wire when it is moved in a magnetic field.	
H8	Calculate the voltage on the secondary of a transformer.	
H9	Calculate energies and potentials of charges in electric fields.	
I1	Work out the charge & energy stored on a capacitor from the charging voltage.	
I2	Work out the capacitance of a network of capacitors.	
I3	Sketch the current / voltage / charge on a capacitor as a function of time as it discharges through a resistor, labelling key points.	
J1	Complete a nuclear equation (including beta decay and neutrinos.)	
J2	Calculate the half life of a radioactive sample from a knowledge of number of nuclei & activity.	
J3	Sketch the no. of nuclei remaining in / activity of a radioactive source as a function of time as it decays, labelling key points.	
J4	Calculate the energy released in a nuclear reaction from the masses of reactants & products OR a graph of binding energy per nucleon.	
K1	Work out galaxy velocity from spectral shift and thus its distance using Hubble's law.	
K2	Perform a variety of exponential calculations.	

Explanation Checklist

Unit	Skill	
L1	How a mass spectrometer works.	
L2	The main categories of fundamental particles.	
L3	The construction of a nuclear (fission) reactor.	
L4	What happens when X-rays hit tissue, and how X-ray images can be improved.	
L5	How an ultrasound image is taken, the significance of acoustic impedance, and the difference between A and B scans.	
L6	How MRI works, and the advantages & disadvantages of MRI in comparison with other techniques (X-ray, PET).	
L7	The life cycle of a star.	
L8	The history of the Universe according to the Big Bang model.	

Chapter A

General Questions

A1 Using and Rearranging Equations

$^9/_{12}$

Use the following equations:

$$s = ut \qquad a = \frac{(v - u)}{t} \qquad F = ma \qquad v = f\lambda$$

$$V = IR \qquad P = IV \qquad E = Pt \qquad Q = It$$

where the letters have the following meanings:

s = distance u, v = velocity t = time m = mass

V = voltage I = current F = force a = acceleration

Q = charge E = energy P = power f = frequency

λ = wavelength R = resistance

A1.1   a) $F = 3.0$ N, $m = 2.0$ kg, what is a?

       b) $I = 0.20$ A, $t = 200$ s, what is Q?

A1.2   Calculate the resistance needed if you want 0.030 A to flow through a component when a 9.0 V battery is connected to it.

A1.3   Calculate the distance travelled by a car going at $30\,\mathrm{m\,s^{-1}}$ in 2.0 minutes.

A1.4   Calculate the wavelength of a wave that travels at $3.0 \times 10^8\ \mathrm{m\,s^{-1}}$ if its frequency is 2.0 GHz (2.0×10^9 Hz).

A1.5   a) Calculate the power of a 0.25 A, 240 V light bulb.

    b) Calculate the power if 5.0 A flows through a 2.0 Ω resistor.

A1.6   A Corsa accelerates from 15 m s^{-1} to 25 m s^{-1} in 8.0 s. Calculate the acceleration.

A1.7   If a jet has a maximum acceleration of 20 m s^{-2}, what is the time it would take to get from 0 m s^{-1} to 100 m s^{-1}?

A1.8   My kettle needs to be able to give 672 000 J of heat energy to water in 240 s. Assuming that it is connected to the 240 V mains, what current is needed?

A1.9   Calculate the force needed if my 750 kg car needs to accelerate from rest to 13 m s^{-1} in 5.0 s.

A1.10 Calculate the electrical energy used by a 240 V light bulb with a resistance of 60 Ω in 600 s.

A2   Derived and Base SI Units

Express the following derived units in terms of the SI base units. The first one has been done for you:

Derived Unit	in Base Units	Power of each base unit			
		m	s	kg	A
m s^{-2}	m s^{-2}	1	-2	0	0
A2.1 J		(a)	(b)	(c)	(d)
A2.2 N		(a)	(b)	(c)	(d)
A2.3 C		(a)	(b)	(c)	(d)
A2.4 V		(a)	(b)	(c)	(d)
A2.5 Ω		(a)	(b)	(c)	(d)
A2.6 Pa		(a)	(b)	(c)	(d)
A2.7 N C^{-1}		(a)	(b)	(c)	(d)
A2.8 V m^{-1}		(a)	(b)	(c)	(d)

Express the following derived units in terms of the unit specified and base units. The first one has been done for you.

A2.9   a) Express the ohm in terms of the volt and base units: $\Omega = V\,A^{-1}$

b) Express the joule in terms of the newton and base unit(s).

c) Express the pascal in terms of the joule and base unit(s).

d) The answer to (c) means that pressure in effect measures an amount of energy per unit _____

e) Express the $V\,m^{-1}$ in terms of the joule and base unit(s).

f) Express the unit of density in newtons and base unit(s).

A3   Standard Form and Prefixes

$^9/_{12}$

You will be penalized if you give the wrong number of significant figures where the question specifies the required number of significant figures. [NOTE: standard form means that there is always one non-zero digit before the decimal point.]

A3.1   Write the following as 'normal' numbers:

a) 3×10^4 b) 4.89×10^6

A3.2   Write the following as 'normal' numbers:

a) 3.21×10^{-3} b) 2×10^0

A3.3   Write the following in standard form to three significant figures:

a) 2 000 000   b) 34 580

A3.4   Write the following in standard form to three significant figures:

a) 23.914   b) 0.000 005 638

A3.5   Write the following as 'normal' numbers with the unit (but without the prefix):

a) 3 kJ   b) 20 mA

A3.6   Write the following using the most appropriate prefixes:

a) 5×10^7 m   b) 6×10^{-10} s

A4   Converting Units

Convert between units as specified. Express your answer in standard form
if the power of ten is ≥ 3, or ≤ -3. Your answer must include units, as
indeed it must in *all* questions with units in this book.

Convert:

A4.1   a) 34.5 mm to nm   b) 34.5 mm to pm

A4.2   2.4 ps to ms

A4.3   a) 465 μA to mA   b) 465 μA to kA

A4.4 43×10^{-7} GW to μW

A4.5   34 m² to cm²

A4.6   58 N m to N cm

A4.7   9600 μm² to cm²

A4.8   0.035 N cm⁻² to Pa

A4.9   450 kg m⁻³ to kg mm⁻³

A5 Gradients and Intercepts of Graphs

$^8/_{10}$

Work out the physical quantity corresponding to the gradient and y-intercept.

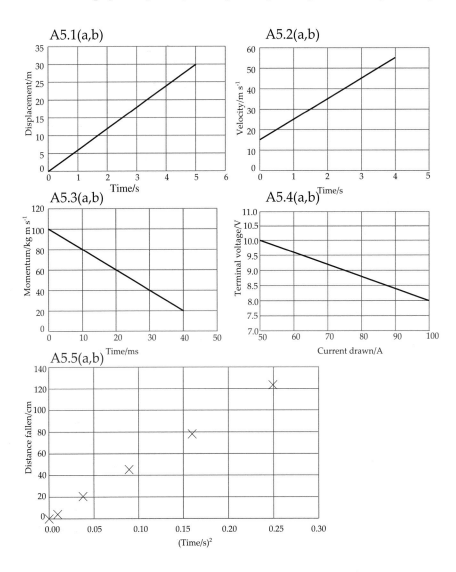

A6 Equations of Graphs

The table shows the formula for a particular physical relationship, and what has been plotted on the x and y axes. You should write down (in terms of the letters in the formula) what the y-intercept and gradient will be. If the graph is not going to be a straight line, write 'not straight' instead of a gradient.

	Equation	y axis	x axis	y-intercept	Gradient
A6.1	$V = \varepsilon - Ir$	V	I	(a)	(b)
A6.2	$s = \frac{1}{2}gt^2$	s	t^2	(a)	(b)
A6.3	$R = \frac{\rho L}{A} + K$	R	L	(a)	(b)
A6.4	$\frac{L}{T} = \lambda f + D$	T^{-1}	λ	(a)	(b)
A6.5	$\frac{1}{R} = \frac{1}{S} + \frac{1}{T}$	R^{-1}	S^{-1}	(a)	(b)
A6.6	$qV = hf - \phi$	V	f	(a)	(b)
A6.7	$d \sin \theta = n\lambda$	$\sin \theta$	n	(a)	(b)

A7   Area Under the Line on a Graph

Estimate the physical quantity corresponding to the area under each line on the graphs (A), (B), (C), (D) and (E).

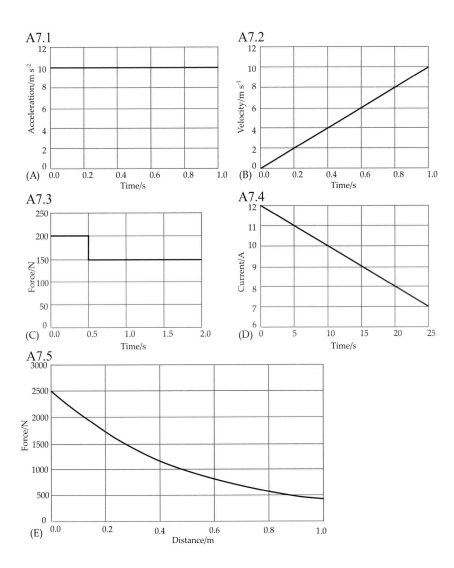

A8   Area Under the Line on a Graph II

Estimate the physical quantity corresponding to the area under each line on
the graphs (A), (B), (C), (D) and (E). You must give your answer in standard
form in SI units, without prefixes. Incorrect unit = no mark.

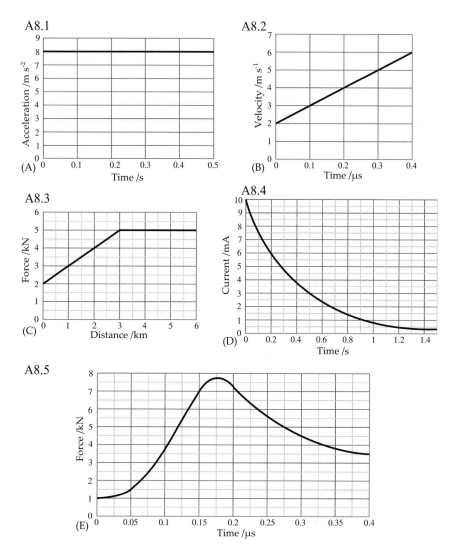

A9   Factor and Percentage Changes

In questions A9.1 to A9.5, give the factor by which the quantity changes to two significant figures. So if it doubles, your answer is 2.0, and if it halves your answer is 0.50.

A9.1   By what factor does $E = mv^2/2$ change if v doubles?

A9.2   By what factor does
   a) $V = -GM/r$ change if r is multiplied by 3.3?
   b) $g = GM/r^2$ change if r is multiplied by 0.64?

A9.3   By what factor does v need to change if $E = mv^2/2$ is to halve?

A9.4   By what factor does d need to change in $I = P/(4\pi d^2)$ if P has multiplied by 5.2 and I is not to change?

A9.5   In $GMT^2 = 4\pi^2 r^3$, by what factor does T change if G and M remain constant, and r is multiplied by 3.5?

In questions A9.6 to A9.10, give the percentage change in the quantity. Use "+" or "−" to indicate increase or decrease (so a 3% decrease would be given as −3%).

A9.6   In $V = IR$, what is the percentage change in V if I increases by 8%?

A9.7   In $E = mv^2/2$, what is the percentage change in E if v increases by 3%

A9.8   In $f = 1/T$, what is the percentage change in f if T increases by 4%?

A9.9   In $r = \sqrt{(A/\pi)}$, what is the % change in r if A decreases by 6%?

A9.10   In
   a) $E = VIt$, what is the percentage change in E if V increases by 1%, I decreases by 2% and t increases by 3%?
   b) $R = \rho L/A$, what is the percentage change in R if L increases by 7% and A increases by 3%?

A10   Proportionality

A10.1 If $V \propto I$ and $I = 0.35$ A when $V = 9.6$ V, what will V be when $I = 0.90$ A?

A10.2 If $E \propto v^2$ and $E = 94$ J when $v = 6.5$ m/s, what will E be when $v = 12$ m/s?

A10.3 If $g \propto 1/r^2$ and $g = 9.8$ N/kg when $r = 6400$ km, what will g be when $r = 15000$ km?

A10.4 If $E \propto x^2$ and $E = 2.5$ J when $x = 1.5$ cm, what will x be when $E = 6.0$ J?

A10.5 If $V \propto 1/r$ and $V = 5000$ V when $r = 7.0$ cm, what will r be when $V = 2000$ V?

A10.6 If $m = \rho a^3$ and $m = 28$ g when $a = 2.5$ cm, what will m be when $a = 8.7$ cm if ρ doesn't change?

A10.7 If $I = P/V$ and $I = 5.2$ A when $V = 230$ V, what will I be when $V = 115$ V if P doesn't change?

A10.8 If $I = P/(4\pi r^2)$ and $I = 6.0$ W/cm² when $r = 3.0$ m, what will r be when $I = 0.30$ W/cm² if P doesn't change?

A10.9 If $R = \rho L/A$, and $R = 5.0$ Ω when $L = 65$ m and $A = 2.5$ mm², what will R be when $L = 120$ m and $A = 1.5$ mm² if ρ doesn't change?

A10.10 If $g = GM/r^2$ and $g = 9.8$ N/kg when $M = 6 \times 10^{24}$ kg and $r = 6400$ km, what will M be if $g = 1.7$ N/kg and $r = 1700$ km if G doesn't change?

Chapter B

Mechanics

B1 Components of a Vector

Where bearings are given, they are in degrees East of North (so North is 000°, East is 090°, South is 180° and West 270°). For the purposes of this exercise, assume that the Earth is flat.

Give the length of the following sides.

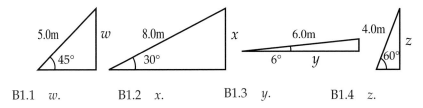

B1.1 w.   **B1.2** x.   **B1.3** y.   **B1.4** z.

B1.5 Eric the Explorer walks 35 km on a bearing of 075°. How far East is he compared to his original position?

B1.6 A trolley has a weight of 11 N and sits on a ramp inclined at 33° to the horizontal. How big is the component of the weight which is trying to pull the trolley along the ramp?

B1.7 A ladder needs to be inclined at 10° to the vertical. It is 6.0 m long, and is propped against a wall. How far will the base of the ladder be from the base of the wall?

B1.8 When walking up Amersham Hill, you walk at an angle of about 6.0° to the horizontal. How far up do you go when walking 500 m along the road?

B1.9 A plumb bob has a weight of 1.0 N. It is swinging on the end of a piece of string and, at one particular instant, the string is inclined at 28° to the vertical. What is the component of the weight perpendicular to the line of the string?

B1.10 A fly in a room is flying on a bearing of 204° at a speed of 0.36 m s^{-1}. Sunlight streams horizontally westward across a room, forming a shadow of the fly on the west wall. How fast does the shadow move?

B2   Adding Vectors

Where bearings are given, they are in degrees East of North (so North is 000°, East is 090°, South is 180° and West 270°). For the purposes of this exercise, assume that the Earth is flat.

B2.1   Work out how far I am from my starting point if I:

   a) walk 3.0 m East then 4.0 m North.

   b) drive 10 km South from the starting point then drive 5.0 km West.

   c) fly 80 km South-West from the starting point then fly 120 km North-West.

B2.2   Work out how fast I am going (relative to a ground-based observer) if:

   a) I row at 9.0 m s^{-1} (relative to the water) South in a river where the water is flowing 1.0 m s^{-1} South.

   b) I swim at 1.0 ms^{-1}(relative to the water) North in a river where the water is flowing 0.30 m s^{-1} East.

   c) in what direction would a ground based observer think I was swimming in question B2.2b? Give your answer as a number of degrees East of North (a bearing).

   d) I fly at 100 km h^{-1} North-West (relative to the air) when the wind is blowing from the North-East at a speed of 20 km h^{-1}.

B2.3   Along which bearing would I have to travel in order to travel North (relative to a stationary observer) if:

   a) I am swimming in a river with a current running 0.40 m s^{-1} to the East, and I can swim at 1.5 m s^{-1} relative to the water?

   b) I am flying in a 15 km h^{-1} wind coming from the West and can fly at 90 km h^{-1} relative to the air?

   c) How fast do I move Northwards over the ground in question B2.3b?

B2.4   A block is subject to two forces - a 90 N force downwards and a 30 N force horizontally to the right.

   a) What is the magnitude of the resultant force on the block?

   b) At what angle, clockwise from the rightward force, does the resultant force on the block act?

B3   Uniform Accelerated Motion in One Dimension

Assume that any dropped or thrown object accelerates downwards at $9.8 \, \mathrm{m \, s^{-2}}$. If a question says that an object is 'dropped', this means that its velocity is zero at the beginning of the motion.

If asked for a velocity or displacement, your answer MUST contain a direction in order to be marked as correct.

B3.1   How far does a dropped pencil case fall in 0.25 s?

B3.2   What is the velocity of a rugby ball 3.0 s after it is kicked upwards with a speed of $16 \, \mathrm{m \, s^{-1}}$?

B3.3   How much time does

    a) a dropped weight take to fall 120 m down a cliff?

    b) the weight take if it were thrown downwards at $2.5 \, \mathrm{m \, s^{-1}}$?

B3.4   A high performance car can travel from rest to $25 \, \mathrm{m \, s^{-1}}$ in 5.0 s.

    a) What is its acceleration?

    b) How far does the car travel while accelerating?

B3.5   An aeroplane cannot take off until it is travelling at $80 \, \mathrm{m \, s^{-1}}$. If its acceleration is $2.5 \, \mathrm{m \, s^{-2}}$, how much distance does it travel while accelerating from rest to its take-off speed?

B3.6   The brakes on a car can stop it from a speed of $31 \, \mathrm{m \, s^{-1}}$ (70 mph) in a distance of 70 m. Calculate the acceleration of the car as it slows down.

B3.7   The Dodonpa roller coaster accelerates from rest to $48 \, \mathrm{m \, s^{-1}}$ (107 mph) with an acceleration of $26.5 \, \mathrm{m \, s^{-2}}$. How much time does it take?

B3.8   A tennis ball is fired upwards at a speed of $60 \, \mathrm{m \, s^{-1}}$ from the top of a tall cliff. Where is it in relation to the starting point after 12.0 s?

B3.9   You want to fire a ball vertically into the air so that it reaches a maximum height of 100 m. How fast must you fire it?

B4   Trajectories

Complete the values in the table below, assuming that all projectiles are launched horizontally and fall downwards with an acceleration of 9.8 ms^{-2}.

	Horizontal distance* /m	Horizontal speed /m s^{-1}	Time to target /s	Distance fallen /m
B4.1	4.0	4.0	(a)	(b)
B4.2	(a)	20.0	(b)	0.020
B4.3	86	220	(a)	(b)
B4.4	1.5	(a)	(b)	0.32
B4.5	(a)	280	(b)	6300

* to target

B4.6   You shoot horizontally at a target that is 30 m away with a gun which fires a bullet at 150 m s^{-1}. How high must the gun be above the target in order to hit it?

B4.7   You are trying to drop essential survival supplies from an aeroplane to help the survivors of a crash who are stranded. You are flying 300 m above them, and your aircraft can travel no slower than 30 m s^{-1}. You fly on a straight line which will pass over the survivors. How far (in metres) in advance of overflying the survivors do you need to drop the package?

B4.8   A rugby player is aiming for a conversion. He kicks the ball at 15 m s^{-1} at an angle of 50° to the horizontal. At the time, he is 20 m from the posts.

a) How long will the ball take to reach the posts?

b) How high will the ball be when it reaches the posts?

B4.9   A cricket batsman hits a ball at a speed of 27 m s^{-1} at an angle of 60° to the horizontal. How far away would you have to stand in order to catch it, assuming you want to catch it just before it hits the ground?

B5 Moments

Strength of Earth's gravity at ground level = 9.8 N kg^{-1}. 1 tonne = 1000 kg.

As throughout this book, numeric answers should contain units. Where forces are asked for, ensure that the direction is in the answer (e.g. up/down). Assume that the mass is evenly distributed in the rulers, poles, planks, bridge spans mentioned in the questions.

B5.1 A metre rule is pivoted about the '50 cm' mark (which is the position of its centre of mass). In each case, find the direction and magnitude of force F needed to balance the rule.

    a) There is a 3.0 N upwards force at the 20 cm mark. Force F acts at the 70 cm mark.

    b) There is a 5.0 N upwards force at the 10 cm mark. Force F acts at the 60 cm mark.

    c) There is a 2.0 N upwards force at the 5.0 cm mark, and a 12 N downwards force at the 40 cm mark. Force F acts at the 75 cm mark.

    d) There is a 100 g mass sitting on the 10 cm mark, and a 50 g mass sitting on the 60 cm mark. Force F acts at the 30 cm mark.

B5.2 A metre stick has its centre of mass at the 50 cm mark, and weighs 0.92 N. A 2.00 N weight is stuck to the 10 cm mark with massless glue. About which point will the ruler balance?

B5.3 A 200 m bridge span is supported at both ends. The span has a mass of 100 tonnes. A 30 tonne bus is 50 m from one end of the span. Calculate the supporting force holding the bridge up at the end nearer the bus.

B5.4 Two workers are moving a 20 kg, 10 m scaffolding pole. One stands at the end, the other stands 2.0 m from the other end. Calculate the force exerted by the worker standing at the end in holding the pole.

B5.5 Calculate the weight of the pole 'carried' by the other worker in question B5.4.

B5.6 Two pupils who don't like each other are made to carry a 1.0 m × 2.0 m whiteboard down some stairs. Each takes their share of the weight by holding the bottom corner at their end. Assuming that they each want the easier job, which end should they fight over?

B5.7 The pub sign shown in the diagram is supported by a hinge and by a
metal rod. Calculate the tension in the rod if the pub sign is an 80 cm
square of mass 30 kg. Ignore the mass of the rod, and assume that the
hinge is well-oiled.

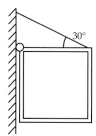

B6 Stress, Strain and Young's Modulus

Preliminary question: A wire has a cross sectional area of 1.5 mm². What is its cross sectional area in m²? Answer: 1.5×10^{-6} m². If you can't see why this is true, ask your teacher before continuing with this section.

B6.1 Complete the questions in the table. All samples have a circular cross section.

Diameter /mm	Cross Sectional Area /m²	Tension /N	Stress /MPa
1.0	(a)	15.0	(b)
3.2	(c)	420	(d)
0.30	(e)	(f)	320

B6.2 A rubber band was 8.4 cm long before it was stretched. It is then stretched until its strain is 2.45. Calculate its new total length.

B6.3 A copper wire is put under tension. It was 1.52 m long to begin with, and then stretches by 3.2 mm. Calculate its strain.

B6.4 A brass pin has a cross sectional area of 0.50 cm². Brass has a tensile strength of 190 MPa. Calculate the maximum tensile force it ought to be able to withstand without breaking.

B6.5 Mild steel has a breaking strength of 500 MPa. If you want to support a 200 kg piano using a single steel wire, what is the minimum diameter of wire you require?

B6.6 A bolt is needed to attach an actor's harness to a wire across a stage. The bolt is 5.0 cm long with a 0.25 cm² cross sectional area, and must extend by no more than 0.02 mm when supporting a 900 kg mass. Calculate the minimum value of Young's modulus of a material if it is to be suitable.

Assuming that all of the materials obey Hooke's Law and that they are circular in cross section, find the specified values in the table:

	Diameter /mm	Cross Sectional Area /m²	Original Length /m	Tension /N	Extension /mm	Stress /MPa	Strain	Young's Modulus /GPa
B6.7	1.0		56	890	32	(a)	(b)	(c)
B6.8			7.36		(a)	500	(b)	211
B6.9	(a)	(b)		9.8		(c)	0.40%	130
B6.10		1.5×10^{-6}	(a)		5.0	70	(b)	130

B7 Springs

Numeric answers should ideally be in SI form, without prefixes. For questions with two parts, both answers must be correct for the mark.

B7.1 A spring with constant 50 N m^{-1} has a load of 12.5 N applied to it. What will be its extension?

B7.2 A spring of natural length 10.0 cm and spring constant 4.00 N cm^{-1} has a load of 22.0 N placed on it. What is its new length?

B7.3 If a spring of natural length 1.50 cm stretches to 1.65 cm when a 16 N force is applied, what is its spring constant?

B7.4 What mass should be suspended from a spring of length 20 cm and spring constant 6.0 kN m^{-1} in order for the spring to be stretched to a length of 22 cm.

B7.5 A spring has natural length of 8.60 m, and you know that it requires a force of 30 N to stretch it to a length of 9.15 m. Work out the spring constant.

B7.6 A spring with natural length 0.70 m requires 3.2 N to stretch it by 17.5 cm. Work out the force required to stretch the spring to a length of 83 cm.

B7.7 Two identical springs, each of natural length 2.0 m and spring constant 80 N m^{-1} are placed in series (that is, one joined to the end of the other), with a weight of 7.5 N suspended from the bottom spring.

   a) State the tension in each spring.

   b) Work out the total extension of the system.

   c) If the two identical springs were placed in parallel so that they can share the load, with the same weight of 7.5 N suspended from the combination, work out the tension in each of the springs.

   d) What is the total length of the system?

B7.8 If three identical springs were put in series, how would:

   a) the spring constant of the system and

   b) the total extension of the system compare to just one of the springs on its own with the same force applied?

B7.9 If five identical springs were placed in parallel, how would:

   a) the stiffness of the system

   b) the total extension of the system compare to just one of the springs on its own with the same force applied?

B8   Work, Energy and Power

B8.1   A box of mass 5.0 kg is dropped from a height of 3.2 m.

    a) How much gravitational potential energy (GPE) was stored by the box before it was released?

    b) The box lands on a table that is 70 cm above the ground. How much work did gravity do on the box on its way down to the table?

B8.2   An object of mass 3.5 kg slips all the way down a slope inclined at 40° to the horizontal, with a base length of 4.8 m.

    a) How much GPE does the object lose?

    b) If the average frictional force is 4.0 N, work out how much work the object does against friction.

B8.3   50 J of work is done in stretching a spring to an extension of 3.5 cm. Work out the average force applied.

B8.4   A boy whirls a 30 g conker around his head in a circle at a speed of 2.2 m s^{-1}, using a taut inextensible string. How much work is done on the conker by the tension in the string?

B8.5   A weight lifter pulls a 2000 kg car forwards at an average speed of 1.5 m s^{-1} against a force of 1250 N.

    a) What is his power output?

    b) How long will it take the weight lifter to do 10 000 J of work?

B8.6   A 1300 kg car travels at a steady speed, covering 75 m in 5 seconds. Frictional forces are constant and are 450 N in total. Work out the power output of the engine, assuming 100% efficiency.

B8.7   A child of 40 kg rides a 35 kg bike at 9.0 m s^{-1}. The brakes are then applied and the bike is slowed to 3.8 m s^{-1}. How much work is done by frictional forces?

B8.8   A 55 kW motor is used to lift a 4800 kg mass vertically up a mine shaft. What is the maximum possible speed that the mass could move upwards?

B8.9   A 4.0 kg ball is thrown vertically up into the air with an initial velocity of 8.5 m s^{-1}. By the time it is height h metres above the starting point, it has a velocity of 3.0 m s^{-1} and has done 4.0 J of work against air resistance. Find h.

B9 Energy, Springs and Materials

Give numeric answers in SI form, without prefixes. Assume extension is proportional to tension in questions B9.1 - B9.5 & B9.7.

B9.1 A spring of natural length 50 cm is stretched to 56 cm when a force of 4.0 N is applied.

    a) How much work is done on the spring to stretch it?

    b) How much elastic potential energy (EPE) is stored in the spring?

B9.2 A spring of natural length 30 cm with spring constant 8.0 N cm^{-1} stretches by an extra 20%. Work out how much elastic potential energy is stored in the spring.

B9.3 A spring with natural length 75 cm requires a force of 300 N in order for it to stretch to 85 cm. How much EPE would be stored in the spring if it were stretched to 90 cm?

B9.4 60 J of work is done to stretch a spring with spring constant 7.5 N cm^{-1} from its natural length of 0.24 m to some new length. Work out this new length.

B9.5 Calculate how much extra work must be done in order to stretch a spring from 17 cm to 20 cm, if its spring constant is 300 N m^{-1} and natural length 15 cm.

B9.6 Estimate the work done in stretching the spring in the graph below:

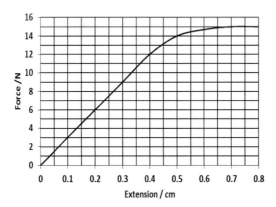

B9.7 A wire of natural length 50 cm, diameter 1.5 mm and Young's modulus 3.2 GPa is stretched to a new length of 52.4 cm, which is below the limit of proportionality. How much work was done in order for this to happen?

Chapter C

Electric Circuits

C1 Combinations of Resistors

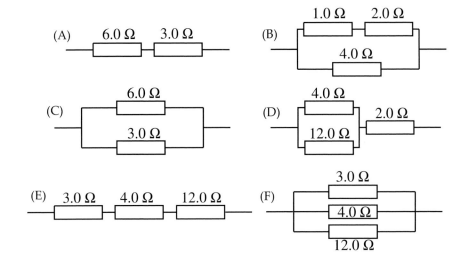

What is the resistance of labelled combination?

C1.1   a) A   b) B

C1.2   a) C   b) D

C1.3   a) E   b) F

Resistivity

Complete the questions in the table:

Length /m	Wire thickness	Resistivity /Ω m	Resistance /Ω
68	cross sectional area: 2.1×10^{-6} m^2	1.5×10^{-8}	C1.4
C1.5	cross sectional area: 0.50×10^{-6} m^2	4.9×10^{-7}	15
1.0	1.0 mm radius	4.9×10^{-7}	C1.6
15000	1.0 cm diameter	1.5×10^{-7}	C1.7

C1.8   Conventional domestic 13 A sockets are connected with copper cables with a cross sectional area of 2.5 mm^2. Copper has a resistivity of 1.5×10^{-8} Ω m. What is the resistance of 20 m of cable?

C1.9   A high voltage wire for transmission of electricity across the country is made of 10 aluminium wires (resistivity $= 2.5 \times 10^{-8}$ Ω m) wound together with 15 copper wires (resistivity of 1.5×10^{-8} Ω m). If all of the wires have a radius of 2.0 mm, calculate the overall resistance of 20 km of cable. (The aluminium is there to give strength to the cable.)

C2   Charge Carriers

Data: Magnitude of the charge on an electron = 1.60×10^{-19} C
Free electron density of copper [Cu] = 10^{29} m^{-3}
Free electron density of germanium [Ge] = 10^{20} m^{-3}

C2.1   How many electrons are needed to carry a charge of -6.00 C?

C2.2   How many electrons flow past a point each second in a 5.0 mA electron beam?

C2.3   Alpha particles have twice the charge of an electron. What is the current caused by a radioactive source which emits 3000 alpha particles per second?

C2.4   An electron gun emits 3.0×10^{21} electrons in two minutes. What is the beam current?

C2.5   Assume all wires have a circular cross section. Calculate the values to complete the gaps in the table:

Diameter /mm	Cross Sectional Area /mm^2	Material	Current /A	Drift Velocity /m s^{-1}
	2.5	Copper	13	(a)
	0.75	Copper	6.0	(b)
1.0		Copper	(c)	0.005
	(d)	Copper	2.0	0.20
(e)		Germanium	2.0	0.20

C2.6   In an experiment, a current of 3.5 A is being passed through a copper sulphate solution in a 10 cm cubical container, with the electrical terminals being opposite faces. This contains equal numbers of Cu^{2+} and SO_4^{2-} ions which have respectively +2 and −2 electron charge units. Assuming that the two ions have equal speed in the solution, and that there are 6.0×10^{26} of each per cubic metre of the solution, work out their mean speed.

C3   Charge Carriers II

Data: Magnitude of the charge on an electron = 1.60×10^{-19} C
Free electron density of copper [Cu] = 10^{29} m^{-3}
Free electron density of germanium [Ge] = 10^{20} m^{-3}

C3.1   If 0.035 pC of charge is transferred via the movement of Al^{3+} ions, how many of these must have been transferred in total?

C3.2   If a 50 μA current is flowing then find out how many electrons pass a point each minute.

C3.3   In a bolt of lightning, 45 nC flows to ground in 25 ms. Work out the average number of charge carriers flowing past per second.

C3.4   If 56×10^{16} electrons flow to the ground in 0.035 μs, work out the average current.

C3.5   How long does it take for a current of 6.0 A to deliver 1.5×10^{17} Cu^{2+} ions in a solution? Assume these ions are the only charged particles moving.

C3.6   In an MgSO$_4$ solution, a current of 36 μA flows. Work out how many SO$_4^{2-}$ ions pass a point in 15 seconds. [Hint: assume that the Mg^{2+} ions and SO$_4^{2-}$ ions move at the same speed in opposite directions and the movement of each type of ion is responsible for half the current]

C3.7   Complete the questions in the table. In all cases, free electrons are travelling in a wire of circular cross section, area A, diameter D.

D /mm	A /mm^2	Material	Current /A	Drift Velocity
	3.8	Ge	7.0	(a)
2.5		Cu	4.0	(b)
1.0		Ge	(c)	0.0050 m s^{-1}
	(d)	Cu	6.0	40 mm s^{-1}
(e)		Ge	2.0	75 mm s^{-1}

C3.8   A copper wire with diameter 0.90 cm has a current of 3.0 A flowing through it. It is connected in series to another wire, identical except that it has radius of 0.15 cm. Work out the ratio of the drift velocity in the thick wire to the drift velocity in the thin wire.

C4　Kirchhoff's Laws

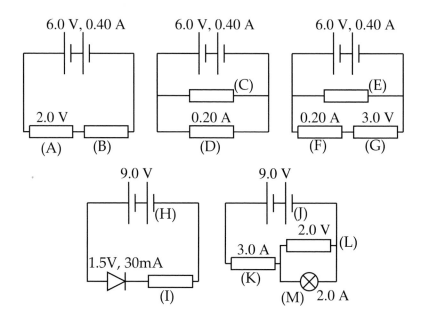

If they are not given, fill out the currents and voltages for the question parts below:

	Current /A	Voltage /V
C4.1	(A) (a); (B) (b)	(A); (2.0) (B) (c)
C4.2	(C) (a); (D) (0.20)	(C) (b); (D) (c)
C4.3	(E) (a); (F) (0.20); (G) (d)	(E) (b); (F) (c); (G) (3.0)
C4.4	(H) (a); (I) (b)	(H) (3.0); (I) (c)
C4.5	(J) (a); (K) (3.0); (L) (c); (M) (2.0)	(J) (9.0); (K) (b); (L) (2.0); (M) (d)

Additional Resistor combinations and power – on-line

isaacphysics.org/assignment/phys19_c4_add

C5   Potential Dividers

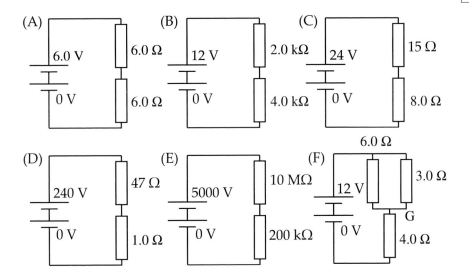

C5.1   What is the voltage across the bottom resistor in circuit (A)?

C5.2   In circuit (B):

    a) What is the voltage across the bottom resistor?

    b) What would the potential of the point between the resistors be if the 2.0 kΩ resistor were removed, leaving a gap in its place?

    c) What would the potential of the point between the resistors be if the 4.0 kΩ resistor were removed, leaving a gap in its place?

    d) What would the potential of the point between the resistors be if the 2.0 kΩ resistor were removed and a wire was attached in its place to complete the circuit?

    e) A voltmeter with resistance 10 kΩ is used to measure the voltage across the 4.0 kΩ resistor. What would it read?

C5.3   What is the voltage across the bottom resistor in circuit (C)?

C5.4   What is the voltage across the bottom resistor in circuit (D)?

C5.5   What is the voltage across the bottom resistor in circuit (E)?

C5.6   What is the potential at G, the junction between the two resistors in parallel and the one in series, in circuit (F)?

C5.7   The 8.0 Ω resistance in circuit (C) is a loudspeaker (the battery represents the amplifier). The other resistor is replaced with a variable resistor which can take all values between 0 Ω and 30 Ω, and is used as a volume control. This volume control changes the voltage across the speaker. What is the range of speaker voltages which are possible? (Give the minimum and maximum.)

C5.8   A thermistor has a resistance of 800 Ω at a temperature of 16 °C. It is wired in series with a fixed resistor and a 9.0 V battery. A high-resistance voltmeter is connected to give a 'temperature' reading.

   a) If the voltage reading is to go up when the temperature increases, should the voltmeter be connected in parallel with the thermistor or the fixed resistor?

   b) If the voltmeter needs to read 3.0 V when the temperature is 16 °C, what is the resistance of the fixed resistor?

C6   Internal Resistance

$^8/_{10}$

C6.1   Give the missing values in the table:

e.m.f /V	Internal Resistance /Ω	Current /A	Terminal p.d. /V	Load Resistance /Ω
12.0	(a)	20	10.2	
12.0	0.12	72	(b)	
230.0	0.53	(c)	227.5	
6.0	(d)		4.2	4.3
(e)	3.2		21.3	12.0

C6.2   A school high voltage power supply unit has an e.m.f. of 5.0 kV. If short circuited, the current must be no more than 5.0 mA. Calculate the internal resistance of the supply needed in order to achieve this.

C6.3   A small battery is powering a powerful lamp. The terminal p.d. is 11.3 V, and the current flowing is 10.2 A. Assuming that the battery has an internal resistance of 2.4 Ω, calculate the e.m.f. of the battery.

C6.4   A high-resistance voltmeter is connected in parallel with a portable battery used to start cars. Before the car is connected, the meter reads 12.4 V. When the car is connected, and a 64 A current is flowing, the meter reads 11.5 V.

a) What is the e.m.f. of the battery?

b) What is the internal resistance of the battery?

C6.5   You are building a power supply which needs to be able to handle currents of zero to 10 A. Assume that you build it to have a terminal p.d. of 13.5 V when disconnected, and 10.5 V when supplying 10 A. (a) State the e.m.f. (b) Calculate the internal resistance of the supply.

Additional Internal Resistance – on-line

isaacphysics.org/assignment/phys19_c6_add

Chapter D

Waves

D1 Amplitude and Intensity

D1.1 5.0 W of light from a lamp shines on a 2.0 m × 3.0 m wall. Calculate the intensity.

D1.2 A 0.50 W laser is shone on a wall, making a circular spot of diameter 1.0 mm. Work out the intensity.

D1.3 Work out the power of the source needed to cover a 7.0 m × 7.0 m stage with light to an intensity of 300 W m^{-2}.

D1.4 If one day the solar intensity incident on a part of England is 400 W m^{-2}, work out the total energy that would arrive in one minute on a square piece of land 2.0 km × 2.0 km.

D1.5 a) One laser emits light that has amplitude 200 V m^{-1} and intensity 0.26 W m^{-2}. Another laser emits light of amplitude 300 V m^{-1}. In all other respects it is identical. Work out its intensity.

b) A third similar laser emits light with intensity 1.5 W m^{-2}. Work out the amplitude of the light.

D1.6 Three sets of ripples on the surface of a pond have amplitudes 1.5 cm, 2.25 cm and 3.0 cm respectively. Work out the ratios of the intensities of these three waves.

D1.7 The light from a bulb shines equally in all directions.

a) If 20 W of light is given off, what will the intensity be 12 m from the lamp? (Consider the shape of the region illuminated if the light hits this surface after travelling 12 m in all directions.)

b) What will the intensity be at a distance of 24 m?

D1.8 The Sun is 1.5×10^{11} m from the Earth. If the power incident on Earth is approximately 1.0 kW m^{-2}, calculate the total power (luminosity) of the Sun.

D2   Polarisation

For each polariser, the angle given is the one for which light is transmitted and is given clockwise from the vertical. Where there are multiple answers required, both must be correct for the mark.

D2.1   Horizontally polarised light is shone on a polariser that is angled at 35° to the vertical. The incoming light has amplitude 200 V m^{-1} and intensity 53 W m^{-2}. Work out (a) the amplitude and (b) intensity of the transmitted light.

D2.2   Unpolarised light of intensity 4.0 W m^{-2} is incident on a polariser placed at 15° from the vertical. State the intensity of the transmitted light.

Vertically polarised light of amplitude 100 V m^{-1} and intensity 14 W m^{-2} is incident on the following combinations of polarisers (P1, then P2, then P3). Complete the values indicated in the table below. The polariser angles P1, P2 and P3 are from the vertical.

	Polariser angle /°:			Transmitted light		
	P1	P2	P3	Amplitude /V m^{-1}	Intensity /W m^{-2}	Angle (to vertical)
D2.3	0	20	0	(a)	(b)	
D2.4	90	35	n/a		(a)	
D2.5	15	50	50	(a)		(b)
D2.6	0	45	90		(a)	(b)
D2.7	15	105	60	(a)		
D2.8	10	165	95	(a)	(b)	(c)

D3   Path Difference

Calculate the gaps in the table. The speed of sound in air is 330 m s^{-1}.

	Wavelength	Path Difference	Phase Difference	Fully Constructive Interference (Y/N)	Fully Destructive Interference (Y/N)
D3.1	320 mm	160 mm	(a)	(b)	(c)
D3.2	320 mm	(a)	0°	(b)	(c)
D3.3	320 mm	(a)	90°	(b)	(c)
D3.4	633 nm	(a)	180°	(b)	(c)
D3.5	3.00 m	31.5 m	(a)	(b)	(c)

D3.6   Signals of wavelength 1.5 μm are travelling down a 10 km optic fibre. Some travel 'straight down the middle', while others 'zig-zag' back and forth in the fibre. It is required that the maximum phase difference thus caused is no more than 30°. Calculate the maximum path difference allowed.

D3.7   Two aerials are 2.50 m apart, and both are receiving the same radio signal with a frequency of 125 MHz.

a) The phase difference between them is measured as 114°. Calculate the path difference between the two aerials.

b) The aerial which receives the radio signal first is directly North of the one which receives the signal slightly later. What are the possible bearings of the transmitter from the receiving aerials? You may assume that the transmitter is many kilometres from the receiving aerials, and therefore that the path of the waves travelling to the two receivers are effectively parallel when measured in the vicinity of the receiving aerials.

NB The method of questions D3.7 is used to locate the origin of distress and emergency signals. The procedure is called VDF, standing for 'Very high frequency Direction Finding'.

D3.8 Two speakers are set up 13.5 m apart in an auditorium, pointing at each other. A pure sound of frequency 256 Hz is being played through them. You may assume that the phase difference of the signals driving the speakers is $0°$.

   a) A person is standing on the line joining the speakers, 0.25 m from the mid point. Calculate the phase difference as it would be detected by the person.

   b) The person moves to the mid point between the speakers (where the sound is loudest due to constructive interference), and then walks towards one speaker until the sound waves cancel out. How far do they walk until they find this point of near silence?

D4   Interference

$^9/_{12}$

D4.1 Complete the questions in the table:

Wavelength	Slit Separation	Distance to Screen /m	Fringe Spacing /mm
633 nm	0.10 mm	4.00	(a)
530 nm	(b)	6.00	4.0
(c)	1.0 mm	1.50	0.20
0.30 cm	0.10 m	2.50	(d)

D4.2 Complete the questions in the table:

Wavelength	Slit Separation	Order of interference n	Angle to 'Straight Through' Direction
633 nm	0.10 mm	2	(a)
530 nm	600 lines/mm	1	(b)
(c)	1000 lines/mm	1	$10°$
1.0×10^{-11} m	(d)	3	$20°$

D4.3   A diffraction grating has 600 lines/mm. Yellow light from a street-
lamp is shone onto the grating. The yellow light contains two main
wavelengths - of 589.6 nm and 589.0 nm. Calculate the angular separation of the second order ($n = 2$) of these two components as they
emerge from the grating.

D4.4   A slide looks like it has one fine transparent line ruled on a black
background. In fact there are two lines very close together. When
red light (633 nm) is shone through it, and a screen is placed 5.0 m
away from the slits, ten fringe-spacings measure 5.3 cm. Calculate the
separation of the slits on the slide.

D4.5   The light from a 'special LED' consists of two colours of light with
wavelengths of 530 nm and 630 nm respectively. The light is shone
through a diffraction grating with 500 lines/mm, and the two colours
need to be separated by at least 5.0°. What is the minimum order of
interference needed in order to do this?

D4.6   A teacher is trying to demonstrate 'Young's fringes' using green (530 nm)
light. Assuming that the slit separation is 0.050 mm, how far away
from the slits will she need to put the screen to ensure that the fringe
spacing is at least 1.0 mm?

D5   Standing Waves

The speed of sound in air is 330 m s^{-1}.

Consider a particle that is at a particular antinode (point A) of a standing wave. Fill in the table below to state how the motion of certain other particles will compare to this one. [For amplitude, state whether it will be smaller/larger/the same; for phase, state the phase difference in degrees.]

	Position of Particle	Amplitude	Difference in Phase
D5.1	At the next antinode along from point A	(a)	(b)
D5.2	Between point A and the next node	(a)	(b)
D5.3	Beyond the next node along from A, but before the next antinode	(a)	(b)

D5.4   What is the difference between the "amplitude" and the "displacement" of a particle at an antinode?

D5.5   Two waves of amplitude 4.0 cm and frequency 14 Hz are moving in opposite directions at 5.6 m s^{-1} along a stretched string.

   a) If a standing wave were formed, how far apart would you expect the antinodes to be from the nodes on either side of them?

   b) If the string had two fixed ends, what is the minimum length it must be in order for a standing wave to be possible?

   c) If the string had two fixed ends and was 0.70 m long, why would no standing wave be formed?

D5.6   A wind instrument is 60 cm long, and can be modelled as a tube with one closed end and one open end.

   a) What is the lowest frequency that can be played?

   b) If a note of the second-lowest possible frequency were played, state the positions of the nodes (measured from the closed end).

c) If a note of the third-lowest possible frequency were played, state the positions of the nodes.

D5.7   Two microwave emitters are placed facing each other about a metre apart and coherently emit microwaves of the same frequency. A detector moved back and forth between them detects regions of maximum intensity spaced 4.0 cm apart. Calculate the frequency of the microwaves.

D5.8   A musical note of several frequencies is sounded at the mouth of a 1.0 m long vertical tube that has some water in the bottom.

a) Give the depth of water in the tube if the fundamental frequency heard is 125 Hz.

b) When the lowest frequency above the fundamental is played, at what height will the particles' displacement be out of phase and have the same amplitude as particles 8.0 cm above the surface of the water? Give your answer as a distance above the surface of the water.

$^{13}/_{16}$ **D6   The Photoelectric Effect**

When the photon energy is insufficient for the ejection of photoelectrons, answer the question by writing 'no electrons emitted'.

D6.1   Complete the questions in the table:

Frequency of Light /Hz	Wavelength of Light /nm	Work Function	Max. KE of Photoelectrons	Stopping Potential /V
6.0×10^{14}		1.2×10^{-19} J	(a)	
6.0×10^{14}		2.6 eV		(b)
	350	2.6 eV		(c)
	530	(d)		1.35

D6.2   A material's work function is 1.3 eV. Calculate its threshold frequency.

D6.3   A material will not emit photoelectrons unless it is irradiated by light
       with a wavelength less than 380 nm. Calculate its work function in
       electronvolts.

D6.4   a) Calculate the maximum speed of the photoelectrons emitted
              when a material with an 8.4×10^{-20} J work function is illuminated
              by light of frequency 7.0×10^{14} Hz.

           b) What is the minimum speed of the photoelectrons emitted?

D6.5   A graph of stopping potential (y) against frequency of light (x) is
       plotted for zinc, and also for aluminium. Without knowing more
       information, answer the following questions:

           a) Are the lines straight or not?

           b) Are the y-intercepts positive, negative or zero?

           c) Are the gradients positive, negative or zero?

           d) Are the gradients of the two lines the same or different?

           e) Are the y-intercepts of the two lines the same or different?

           f) What is the significance of the x-intercept?

           g) If you answered 'same' to parts (d) or (e), write down the value
              of the common gradient or intercept.

D6.6   A material has a work function of 3.4 eV, and is illuminated by 5.0 eV
       photons. Calculate the stopping potential of its photoelectrons.

D7   Quantum Calculations

D7.1   Complete the questions in the table:

Frequency of Light /Hz	Wavelength of Light /nm	Photon Energy /J	Photon Energy /eV
6.0×10^{14}			(a)
	(b)		1.5
	(c)		2.5
	500		(d)
	1013		(e)
2.0×10^{15}			(f)

D7.2   A laser diode requires 3.2 V across it to make it work. This means that its photons will have an energy of 3.2 eV. Calculate the wavelength of the light emitted.

D7.3   When an electron annihilates a positron, two photons are produced, each with an energy of 511 keV. Calculate the photon frequency.

Caution - when working with particles, do not use $c = f\lambda$. Question D7.9(b) shows you why.

Complete the questions in the table:

	Wavelength /nm	Particle	Momentum /kg m s^{-1}	Kinetic Energy /eV
D7.4	3.0	Electron	(a)	(b)
D7.5	3.0	Neutron	(a)	(b)
D7.6	(a)	Electron ($\beta-$)	(b)	10^5

D7.7   a) Calculate the momentum of an electron if its kinetic energy is 10 keV.

         b) An electron's wavelength is 3.0×10^{-7} m. What is its momentum?

D7.8   The tandem electrostatic accelerator can accelerate carbon-12 nuclei to a kinetic energy of 60 MeV. How fast are they going? Assume $m = 12 \times m_{neutron}$

D7.9   An electron is travelling at 2.0×10^6 m s^{-1}.

         a) Calculate its momentum and its kinetic energy. Now use the momentum to calculate its wavelength and use the energy to calculate its frequency.

         b) Use $c = f\lambda$ to 'calculate' the speed of the electron using the frequency and wavelength. (Extension - what do you notice? Can you spot what has caused this oddity?)

D8   Refraction and Total Internal Reflection

Data: Refractive index of crown glass = 1.51
       Refractive index of flint glass = 1.61
       Refractive index of water = 1.34
       Refractive index of cubic zirconia = 2.16
       Refractive index of diamond = 2.42
       Take the refractive index of air to be 1.00

Complete the table to show the missing angles. In some cases, refraction is
impossible. In these cases give your answer as 'Total Internal Reflection'.

	Light passing from...		...to	
	Material	Angle of Incidence /°	Material	Angle of Refraction /°
D8.1	Air	30	Crown Glass	(a)
	Air	30	Flint Glass	(b)
	Air	13	Flint Glass	(c)
	Air	(d)	Crown Glass	30
D8.2	Crown Glass	50	Air	(a)
	Crown Glass	40	Water	(b)
	Crown Glass	50	Flint Glass	(c)
D8.3	Water	(a)	Air	60
	Flint Glass	(b)	Air	90

Complete the table to show the missing critical angles.

	Boundary between		Critical Angle
D8.4	Water	Air	(a)
	Crown Glass	Air	(b)
	Flint Glass	Air	(c)
	Cubic Zirconia	Air	(d)
	Diamond	Air	(e)
D8.5	Flint Glass	Water	(a)
	Crown Glass	Water	(b)

D8.6   Calculate the speed of light in:

    a) Flint glass.

    b) Diamond.

D8.7   Calculate the speed of light in:

    a) Cubic zirconia as a fraction of the speed of light in air.

    b) Diamond as a fraction of the speed in cubic zirconia.

D8.8   The critical angle for light passing from flint glass into ethanol is 57.6°. Calculate the refractive index of ethanol.

D8.9   When light passes from water into ice at an incident angle of 38.0°, the angle of refraction is 39.0°. Calculate the refractive index of ice.

D9   Atomic Spectra

Questions D9.1 to D9.5 concern an atom with energy level diagram below

Ionization 0 eV
. . . .
————————————-1.6
————————————-4.8

———————————-8.0

———————————-15.0

D9.1   What energy (in eV) is needed to ionize it when in its ground state?

D9.2   What wavelength of light would be emitted if an electron descended from the −4.8 eV state to the ground state?

D9.3   If the atom is in the −1.6 eV state, and the electron descends to the ground state in three separate stages, what is the wavelength of the least energetic photon emitted?

D9.4   When white light is shone onto a gaseous sample of these atoms, which wavelength will be absorbed as atoms excite from the ground state to the first excited state?

D9.5   How much energy (in eV) is the atom likely to absorb:

   a) If a 10 eV photon passes through it in its ground state.

   b) If instead the atom starts in the −8.0 eV state. You should give a different answer to part (a) if at all possible.

   c) If a 10 eV *electron* passes through the atom in its ground state.

D9.6 to D9.8 concern atomic hydrogen, with energy levels (in electron volts, with respect to the ionization level) $E = -13.6 \text{ eV}/n^2$, where n = 1 is the ground state, n = 2 is the first excited state etc.

D9.6   a) Give the energy (in eV) of the photon absorbed when the atom is excited from n = 1 to n = 2.

   b) Give the wavelength (in nm) of the photon emitted when the atom de-excites from n = 3 to n = 2.

D9.7   How much energy is the atom likely to take from:

   a) An 11.0 eV photon passing through it?

   b) An 11.0 eV *electron* passing through it?

D9.8   Would the atom be able to absorb a 22 eV photon?

Chapter E

Uncertainties

E1 Absolute Uncertainties

For the following measured quantities give (A) the absolute uncertainty (including units), (B) the heading you would put in the results table and (C) the number of decimal places you would give in the results table.

Fill in the gaps in the table.

	Quantity being measured	A	B	C
E1.1	Voltage using a multimeter reading in volts to 3 d.p.	(a)	(b)	(c)
E1.2	Current, using a multimeter, reading in amps to 3 d.p. but the last digit fluctuates a lot. The second to last digit stays fairly steady.	(a)	(b)	(c)
E1.3	A length measured using a metre rule that has mm marks.	(a)	(b)	(c)
E1.4	A time, where you are manually operating a stopwatch that reads to the nearest hundredth of a second.	(a)	(b)	(c)

E1.5   You measure the time taken for a pendulum to complete 20 full swings, using an electronic timer accurate to the nearest 0.1 s. You then divide your answer by 20 to get the time for just one swing. What is the absolute uncertainty on your value for just one swing?

43

Now work out the absolute uncertainties in your readings if you obtained the following data after conducting repeats:

E1.6 For a current, where your ammeter can read to the nearest 0.01 A, you obtain the results: 2.35 A, 2.39 A, 2.39 A, 2.38 A.

E1.7 Using a micrometer that reads the diameter of a wire to the nearest 0.01 mm, you obtain the results: 3.46 mm, 3.55 mm, 3.42 mm. What is the absolute uncertainty in the radius of the wire measured?

E1.8 If you measured the length of a piece of string and got readings of 6.74 m, 6.79 m, 6.75 m, what would be the absolute uncertainty in the total length of two identical pieces of string placed end to end [assume that the join between them is exact - i.e. no extra uncertainty due to this].

E1.9 If you measured a resistance using an ohmmeter and obtained the following results, give a value for the absolute uncertainty and the average that you would use. 10.5 Ω, 10.3 Ω, 10.9 Ω, 14.7 Ω, 10.6 Ω.

E2   Relative Uncertainties

As elsewhere in this book, give an appropriate number of significant figures (e.g. giving an uncertainty to 3sf, or giving a measurement to 2dp
if the uncertainty is ± 0.1, would be wrong). Please make sure that the
unit of absolute uncertainty is clear - so 20.34 mA ± 20 μA or (20.34 ± 0.02)
mA are both appropriate, but 20.34 mA ± 20 would not be clear. Note that
'nearest millimetre' implies an absolute uncertainty of ± 0.5 mm not ± 1 mm.

E2.1   Calculate the relative uncertainty, in percent, of:

    a) A length of 50.4 cm measured using a metre rule to ± 0.5 mm.

    b) A current of 240 mA measured to the nearest milliamp.

    c) A time of 0.62 s measured using a stopwatch to the nearest 0.01 s.

    d) An angle of 43° measured to the nearest degree with a protractor.

    e) A time of 4 minutes 32 seconds measured to the nearest second.

E2.2   Write the following measurements using an absolute uncertainty with
an appropriate number of significant figures (e.g. as 12 mA ± 1 mA).

    a) A time of 97.35 seconds measured to ± 0.1%.

    b) A voltage of 1.629 V measured to ± 5%.

E2.3   What is the relative uncertainty of a frequency of 20 MHz (exactly)
measured to the nearest 10 kHz?

E2.4   Give the relative uncertainty required in a clock which is put right at
noon on Sunday, and by the following Sunday noon must have an
error of no more than 5 seconds.

E2.5   What is the percentage inaccuracy in a measurement of the speed of
light (which is 3.00×10^8 m s^{-1}) which comes out as 2.76×10^8 m s^{-1}?

E2.6   An experiment is conducted to find the acceleration of a dropped
object (which should be 9.81 m s^{-2}). The measurement obtained is
9.62 m s^{-2} ± 1.5%. Is the experiment accurate?

E2.7   A car should have a braking distance at 30.0 mph of 15.0 m ± 3% or
less. What is the minimum measured braking distance which would
lead to the car failing the test?

E3   Propagating Uncertainties

You will be penalized for an inappropriate number of significant figures (e.g. giving an uncertainty to 3sf, or giving a measurement to 2dp if the uncertainty is ± 0.1). Please make sure that the unit of absolute uncertainties is clear - so 20.34 mA ± 20 μA or (20.34 ± 0.02) mA are both appropriate, but 20.34 mA ± 20 would not be clear. Note that 'nearest millimetre' implies an absolute uncertainty of ± 0.5 mm not ± 1 mm.

Calculate the relative uncertainty, in percent, of:

E3.1   A resistance which is worked out from a voltage known to 3% and a current known to 7%. (Equation: $R = \frac{V}{I}$)

E3.2   A frequency which is worked out from a time period known to 2%. (Equation: $f = \frac{1}{T}$)

E3.3   The density of a cuboid block of iron whose lengths are known to 2% and whose mass is known to 0.1%. (Equation: density = $\frac{\text{mass}}{\text{volume}}$)

E3.4   The time taken for a marble to fall by a distance known to 4%. (Equation: distance = $\frac{1}{2}gt^2$)

E3.5   The resistivity of a constantan wire if the resistance is known to 8%, the diameter to 2% and the length to 5%. (Equation: resistivity = $\frac{RA}{L}$, where A is the cross sectional area)

E3.6   Give the speed (with relative uncertainty) of a car which travels (20.0 ± 0.1)m in (1.3 ± 0.1)s.

E3.7   Give the frequency (with absolute uncertainty) of a wave which travels at (320 ± 15)ms^{-1} and has a wavelength of (32.2 ± 0.3)cm. (Equation: frequency = $\frac{\text{speed}}{\text{wavelength}}$)

E3.8   Two resistors, each of 6 Ω resistance (± 2%) are connected in series. What is the relative uncertainty of the total resistance? (Total resistance of resistors in series = sum of the resistances.)

E3.9   I need to put 3.0 kg of flour in a bowl for making some bread. My scales (which read to the nearest 5 g) only go up to 1.0 kg, so I measure out three equal helpings of flour separately, then put them in the bowl. What is the absolute uncertainty of the mass of flour in the bowl afterwards?

E3.10 The speed of a trolley before and after (1.7 ± 0.1)s of acceleration is measured. Before the acceleration, the trolley moved (100 ± 1)mm in (1.78 ± 0.01)s; after the acceleration it moved the same distance in (0.74 ± 0.01)s. Calculate the relative uncertainty of the measured acceleration. $\left(\text{Equation: Acceleration} = \dfrac{\text{change in velocity}}{\text{time of acceleration}}\right)$

E4   Accuracy, Percentage Difference and Reliability

You will be penalized for an inappropriate number of significant figures (e.g. giving an uncertainty to 3sf) and you must include the relevant unit.
 Data (accepted values):

$g = 9.81$ m s^{-2}
$\epsilon_0 = 8.85 \times 10^{-12}$ F m^{-1}
$G = 6.67 \times 10^{-11}$ N m^2 kg^{-2}

E4.1   In an experiment, you obtain the result that $g = (9.89 \pm 0.06)$ m s^{-2}.

a) What is the percentage difference of your experiment compared to the accepted value? [Note - sometimes this is referred to as the percentage error.

b) Does the accepted value lie within your error bars?

c) Without calculation, will the percentage difference in (a) be greater or less than the percentage uncertainty on your result?

E4.2   Your classmate obtains the result that $g = 7.4$ m s^{-2}, and states that his relative uncertainty was 35%. Work out the upper and lower bounds on his value, and state whether the accepted value lies within his error bars.

E4.3   You do an experiment and work out that the percentage difference between your result and the accepted value is smaller than your relative uncertainty. Does the accepted value lie within your error bars?

E4.4   You measure the resistance of a component as 35.7 m$\Omega \pm 300\mu\,\Omega$. You are told that the percentage difference between your answer and the true value is 0.6%. Work out your percentage uncertainty and hence state whether the accepted value lies within your error bars.

E4.5   You obtain a measurement that $G = 6.95 \times 10^{-11}$ N m^2 kg^{-2}, with a relative uncertainty of 5%. Determine whether or not your experiment is consistent with the accepted value of G.

E4.6   You obtain the following results for the time period of a pendulum: $(561, 563, 569, 562, 565)$ ns. None of these results are anomalous. You are then told that the accepted value is 560.5 ns. Does this lie within your error bars?

E4.7   You conduct an experiment to measure the value of ϵ_0. You work out that the total uncertainty in your experiment is 11%, and that the accepted value of ϵ_0 only just lies within your error bars. You tell your friend all this information, and ask him to work out what result you got. What two possible answers should he give you?

Chapter F

Mechanics

F1   Force and Momentum

In these questions ignore the effects of friction & drag.

F1.1   What is the momentum of a 750 kg car going at 31 m s^{-1}?

F1.2   What is the momentum of an electron (mass = 9.1×10^{-31} kg) travelling at 3.0×10^7 m s^{-1}?

F1.3   If a 20 000 kg bus accelerates from 10 m s^{-1} to 25 m s^{-1}, what is the change in momentum?

F1.4   A 50 g ball is travelling at 2.0 m s^{-1} when it hits a wall and rebounds at 1.5 m s^{-1}. Calculate the change in momentum.

F1.5   A 750 kg car takes 15.3 s to accelerate from 5.0 m s^{-1} to 31 m s^{-1}. Calculate the force needed to do this.

F1.6   A 70 kg person jumps in the air and is travelling downwards at 2.0 m s^{-1} when their feet touch the ground. If it takes the person 0.30 s to stop, calculate the resultant force on them.

F1.7   I am trying to push start a car which has stopped. If the biggest force with which I can push the car is 420 N, and the car has a mass of 1025 kg, how fast will it be going after 8.0 s of pushing?

F1.8   Calculate the force needed to accelerate a 50 000 kg spacecraft from rest to 7000 m s^{-1} in four minutes.

F1.9   An alpha particle (mass = 6.7×10^{-27} kg) is fired at the nucleus in a gold atom with a speed of 3.5×10^6 m s^{-1}. It bounces off at the same speed in the opposite direction. If the collision takes 10^{-19} s, what is the average force?

F1.10 How long would it take a 637 N force to accelerate a 65 kg physics teacher from rest up to a speed of 100 m s^{-1}? (NB this is over 200 mph)

F2   Conservation of Momentum

$^8/_{10}$

F2.1   Two masses, called Alfie and Beth, collide and stick together under four different circumstances, as shown in the four rows of the table below. Calculate the missing measurements:

Before collision				After collision
Alfie's mass /kg	Alfie's velocity /m s^{-1}	Beth's mass /kg	Beth's velocity /m s^{-1}	Velocity /m s^{-1}
30	+2.0	40	+1.5	(a)
60	-1.4	30	+2.8	(b)
120	+1.5	80	(c)	0.0
120	+3.0	(d)	-31	+2.0

F2.2   Charlie is driving her 20 000 kg bus. She stops at a roundabout. Percy is driving his 750 kg Corsa at 15 m s^{-1} behind her. He fails to stop and rams into the back of the bus, sticking to it. The impact releases the brakes on the bus. How fast will the smashed up wreck be travelling immediately after the collision?

F2.3   A neutron (mass = 1 u) is moving at 300 m s^{-1} when it smacks into a stationary ^{235}U nucleus (mass = 235 u), and sticks to it. What will the velocity of the combined particle be?

F2.4   A 7.90 g bullet is travelling at 200 m s^{-1}. It hits a 3.00 kg sack of sand which is hanging by a rope from the ceiling. The bullet goes into the sack, and is stopped inside it by friction with the sand. How fast is the sack going immediately after the bullet has "stopped" inside it[1]? NB you must give your answer to 3 significant figures to be awarded the mark.

F2.5   A rocket (containing a space probe) is travelling at 7000 m s^{-1} in outer space. The 2000 kg probe is ejected from the front of the rocket (forwards) using a big spring. If the speed of the probe afterwards is 7200 m s^{-1}, and the rest of the rocket has a mass of 6000 kg, what is the speed of the rest of the rocket?

---

[1]"stopped" means stopped relative to the sand, not stopped relative to a stationary observer.

F2.6   In a strange form of billiards, the cue ball is one third the mass of the other balls, which are stripey. There is no spin, and I hit a stripey ball centrally with the cue ball (travelling at 1.4 m s^{-1}) such that the cue ball rebounds in the opposite direction with half of its initial speed. What is the speed of the stripey ball?

F2.7   I am stranded, stationary, in space, but near to my spacecraft. I detach my 30 kg oxygen cylinder, and fling it away from the spacecraft with a speed of 3.0 m s^{-1}. If my mass (without the cylinder) is 80 kg, how fast will I travel in the other direction towards my spacecraft?

$^{18}/_{20}$

F3   Units of Rotary Motion

F3.1   How big is 3 rad, when expressed in degrees?

F3.2   How many radians are there in 90°?

Complete the questions in the table by converting the units:

	Time period /s	Frequency /Hz	Angular Velocity /rad s^{-1}	Revolutions per Minute (rpm)
F3.3	0.50	(a)	(b)	(c)
F3.4	(a)	(b)	3.0	(c)
F3.5	(a)	(b)	(c)	3800
F3.6	(a)	50	(b)	(c)
F3.7	2700	(a)	(b)	(c)

F3.8   A car travels 10 km. One of its wheels has a radius of 30 cm. Calculate the angle the wheel turns as the car travels this distance (answer in radians).

F3.9   An astronaut's training centrifuge has a radius of 4.0 m. If it goes round once every 2.5 s, calculate the velocity of the end of the centrifuge arm (4.0 m from the pivot).

F3.10 My washing machine has a spin speed of 1200 rpm, and a drum radius of 20 cm. Calculate how fast clothes go, when up against the side of the drum.

F4   Centripetal Acceleration

F4.1   Complete the questions in the table:

Speed /m s^{-1}	Radius /m	Angular Velocity /rad s^{-1}	Centripetal Acceleration /m s^{-2}
	0.32	5.2	(a)
2.1	0.070		(b)
(c)	30.0		9.8
	(d)	0.20	9.8
60	1200		(e)

F4.2   A car goes round a roundabout at 30 mph (13.4 m s^{-1}) on a circular path with a radius of 8.0 m. Calculate the centripetal acceleration.

F4.3   Calculate the force needed to hold a 55 kg teenager in place when in a horizontal fairground ride of radius 3.5 m going round once in 5.0 s

F4.4   a) Calculate the force needed to keep the Earth (mass 6.4×10^{24} kg) in its orbit around the Sun (radius 1.5×10^{11} m). The Earth takes $365\frac{1}{4}$ days to orbit the Sun once.

   b) What is the name of the force which keeps the Moon in orbit round the Earth?

F4.5   A space station with an 8.0 m radius is spun to give the astronauts something which feels like gravity. If the centripetal acceleration is 9.8 m s^{-2}, calculate the speed at which the walls rotate (in m s^{-1}).

F4.6   Calculate the centripetal force experienced by a 500 g pair of wet trousers when in the spin cycle of a washing machine with a 20 cm drum radius if it rotates at 1200 rpm.

F5   Newtonian Gravity

Fields

F5.1   Complete the questions in the table; you may assume that all measurements are made above the surface of the astronomical body:

Mass of body /kg	Distance from centre of body	Gravitational field strength at this distance /N kg^{-1}
(a)	6400 km = Earth radius	9.8
Earth mass	2 × Earth radius	(b)
4.8×10^8 (asteroid)	6100 km	(c)
(d)	3.2×10^6 m	4.0

F5.2   Calculate the force of attraction between two metal spheres each of mass 20 kg whose centres are 20 cm apart.

F5.3   a) At a distance of 1.0×10^7 m from the centre of planet Mogg, the gravitational field strength (g) due to Mogg is 2.1 N kg^{-1}. Calculate g at a distance of 5.0×10^7 m.

   b) The planet Mogg is completely spherical, with radius 2.3×10^6 m. Calculate g at a height of 100 km above the surface of the planet. Use the information given in (a).

   c) Using the information in (a) and (b), calculate the gravitational field strength due to planet Mogg at a distance of 3.0×10^6 m from the centre.

Potential

F5.4   For a planet of mass 6.0×10^{24} kg, calculate the following:

   a) The gravitational potential in J kg^{-1} at a distance 6.4×10^6 m from the centre of the planet

   b) The distance from the centre of the planet where the gravitational potential is -1.1×10^6 J kg^{-1}

F5.5   For a planet of mass 1.0×10^{24} kg, calculate the gravitational potential, in J kg^{-1}, at the following distances from the centre of the planet:

    a) (i) 2.0×10^7 m   (ii) 4.0×10^7 m

    b) Calculate the gravitational potential energy of a 200 kg satellite at the point mentioned in (a)(ii).

F5.6   a) Calculate the mass of a star which gives a gravitational potential of -8.9×10^8 J kg^{-1} a distance 1.5×10^{11} m from it.

    b) Calculate the gravitational potential energy of a 6.8×10^{24} kg planet at this point.

F5.7   The gravitational potential at the surface of a moon is -2.8 MJ kg^{-1}, and its radius is 1700 km.

    a) Calculate the potential at a point 3400 km *above the surface* of the moon.

    b) Calculate the height *above the surface* of a point with potential -1.2 MJ/kg.

F5.8   A 2.400×10^{22} kg moon orbits a 7.200×10^{24} kg planet with an orbital radius of 2.500×10^8 m.

    a) Calculate the gravitational potential at the point half way between the centres of the planet and its moon. [For this question and part (b), take the universal gravitational constant to be $G = 6.674 \times 10^{-11}$ N m^2 kg^{-2}.]

    b) Calculate the gravitational potential at a point 6.800×10^8 m from the centre of the planet and on the same side of the planet as its moon.

F5.9   Calculate the escape velocity from the surface of the Earth. The Earth's radius is 6400 km, and its mass was calculated in F5.1(a).

F5.10 Calculate the minimum velocity which a space probe needs to be given to escape from the gravitational field of a star if it starts 1.5×10^{11} m from the centre of the star. The mass of the star is 3.3×10^{30} kg.

F6   Gravity and Orbits

Mass of Earth = 5.98×10^{24} kg
Radius of Earth = 6400 km

F6.1   The Earth takes a year to go round the Sun on an orbit with a radius of 1.50×10^{11} m. Calculate the Sun's mass.

F6.2   Calculate the orbital time period for a planet going round the Sun in an orbit of radius twice that of Earth.

F6.3   Calculate the height above the surface of the Earth of an orbit with a 24 hour time period.

F6.4   The Moon's orbit round the Earth has a radius of 3.8×10^8 m. Calculate the Moon's speed in its orbit.

F6.5   a) If you want something to orbit the Earth at a height of 200 km above the surface, at what speed must it travel?

   b) What is the time period of the orbit?

F6.6   a) Starting from $F = \frac{GMm}{r^2}$ and $F = \frac{mv^2}{r}$, derive Kepler's 3rd Law relating the radius of an orbit r to the mass of the planet M and the orbital speed v.

   b) Repeating this analysis, relate r and M to the period T.

F7   Oscillators

F7.1   A mass on a spring oscillates 5 times in 4.2 s.

   a) Calculate the angular frequency.

   b) Calculate the spring constant if the mass is 300 g.

F7.2   Calculate the maximum speed of an oscillator if its amplitude is 3.0 cm and its time period is 0.65 s.

F7.3   Calculate the maximum acceleration of an oscillator if its maximum speed is 1.2 m s^{-1} and its amplitude is 6.0 cm.

F7.4   A mass of 2.0 kg is suspended from a spring with constant 24 N m^{-1}. Calculate the time period of the oscillation.

F7.5   The height of the water on a beach can be approximated as simple harmonic motion with a period of 12 hours. If the mean water height is 3.5 m, the amplitude of the tide is 1.6 m, and 'high water' occurs at 7am one day, what would you predict the height of the water to be at 11am?

F7.6   A man jumps off a bridge attached to a bungee. The time period of the oscillation is 4.7 s, and its amplitude is 6.2 m.

   a) Calculate his maximum speed as he goes up and down.

   b) The man has a mass of 85 kg. Calculate the maximum resultant force acting on him during the motion.

   c) Calculate the 'spring constant' of the bungee rope using the information given.

F7.7   Dr Nasty hates laundry. He designs 40 kg washing machines which resonate when they spin the clothes. His machine spins at 1200 rpm, and when it resonates, it lurches about in the kitchen, putting holes in the cupboards and making a lot of noise. Calculate the 'spring constant' he designs the machines to have in order to achieve his horrible plan.

Chapter G

Gases and Thermal Physics

$^7/_8$ **G1   Kelvin Scale of Temperature**

Convert the following temperatures into the 'other' unit. Remember that $0\,°C = 273\,K$ (no $°$ in K).

G1.1   23 °C

G1.2   90 K

G1.3   4 K

G1.4   300 K

G1.5   600 °C

G1.6   −90 °C

G1.7   37 °C

G1.8 $1.5 \times 10^7\,°C$

$^8/_{10}$ **G2   Gas Laws**

Don't forget that one mole of gas contains 6.02×10^{23} molecules, and that the mass of this number of molecules is called the 'molar mass'. Take the gases to be ideal.

G2.1   What is the volume of a mole of gas at atmospheric pressure ($1.01 \times 10^5\,Pa$) and at 20 °C?

G2.2   Calculate the density of nitrogen gas at atmospheric pressure and at 20 °C if the molar mass of nitrogen is 0.028 kg.

G2.3   How many molecules of gas do you need in a 100 cm^3 cylinder to exert a pressure of $1.0 \times 10^8\,Pa$ at a temperature of 800 °C?

G2.4   In the table fill out the missing initial ('$_1$') or final ('$_2$') values:

P_1 /Pa	V_1 /cm^3	T_1 /K	P_2 /Pa	V_2 /cm^3	T_2 /K
1.01×10^5	30	300	(a)	20	300
1.01×10^5	30	300	(b)	30	373
1.01×10^7	2	600	1.01×10^5	(c)	300
1.01×10^5	500	(d)	1.01×10^7	10	4

G2.5   A tyre contains 800 cm^3 of air at a pressure of about 5.0×10^5 Pa at 9.0 °C. After a cycle ride, the volume is 810 cm^3 and the temperature is now 25 °C. Assuming that none of the gas has leaked, what is the new pressure?

G2.6   A tyre contains 800 cm^3 of air at a pressure of about 5.0×10^5 Pa at 9.0 °C. After a cycle ride, the volume is 760 cm^3, the temperature is now 25 °C, and the pressure is 4.0×10^5 Pa. What percentage of the gas molecules have leaked out?

G2.7   A water fire extinguisher contains 4.0 litres of air at 10^7 Pa and 20 °C. When the extinguisher is used, this gas forces the water out. Calculate the pressure when the volume has increased to 10 litres and the temperature has dropped to 3.0 °C.

G3   Heat Capacity

Data: Specific heat capacity of water = 4180 J kg^{-1} K^{-1}
    Specific heat capacity of aluminium = 880 J kg^{-1} K^{-1}
    Specific heat capacity of iron = 435 J kg^{-1} K^{-1}
    Specific heat capacity of paraffin = 2130 J kg^{-1} K^{-1}

$^{10}/_{13}$

G3.1   Complete the values in the table :

Energy /J	Material	Mass /kg	Initial Temperature	Final Temperature
(a)	Aluminium	0.290	15 °C	82 °C
45 200	Paraffin	2.30	3.0 °C	(b)
81 000	Water	1.50	11 °C	(c)

G3.2 How much time will it take a 2300 W kettle to heat 2.31 kg of water from 12 °C to 100 °C? Assume no heat is lost to the surroundings.

G3.3 How much water can a shower head heat each second from 12 °C to 41 °C if the heater has a power of 4200 W? Assume that no heat is lost to the surroundings, and give your answer in kilograms.

G3.4 If 0.024 kg of water gets trapped in the shower heater of question G3.3, the thermal sensor must stop the current before the water reaches 80 °C. Assuming that the water is at 35 °C when the fault occurs, how quickly must the thermal sensor act?

G3.5 A radiator is made using 5.4 kg of iron. It is then filled with 7.3 kg of water. Calculate its heat capacity, that is the heat required to raise the temperature of the whole thing by 1.0 °C.

G3.6 In the following questions, two substances are mixed. In each case work out the missing mass m or temperature t, assuming no heat is lost to the surroundings.

First substance			Second substance			Mixture
Material	m_1 /kg	t_1 /°C	Material	m_2 /kg	t_2 /°C	t_{mix} /°C
Water	3.2	83	Paraffin	4.3	18	(a)
Water	0.34	14	Iron	0.15	230	(b)
Water	1.25	56	Paraffin	(c)	170	84
Aluminium	3.2	12	Paraffin	2.1	(d)	51

G3.7 How much water at 52 °C must I add to 19 kg of water at 21 °C to make it the right temperature, 37 °C for me to bath a baby?

G3.8 If I add 210 g of rivets, made of some unknown metal, at 303 °C to 500 g of water at 15 °C, and the final temperature is 34 °C, what is the specific heat capacity of the mystery metal?

G4   Latent Heat and Heat Capacity

Data: Specific heat capacity of water = 4180 J kg^{-1} K^{-1}
      Specific heat capacity of ice = 2030 J kg^{-1} K^{-1}
      Specific latent heat of fusion of ice = 3.35 × 10^5 J kg^{-1}
      Specific latent heat of vaporization of water = 2.26 × 10^6 J kg^{-1}
Assume that the given heat capacities remain constant with temperature.

G4.1   A frozen pipe contains 5.60 kg of ice.

  a) How much energy is needed to melt it without changing its temperature?

  b) If, in fact, the ice were initially at −3.5 °C, how much energy would be taken to warm it to melting point and then melt it?

G4.2   A certain quantity of ice requires 10.0 J to warm it to melting temperature. It then requires 100 J to melt it.

  a) Calculate the initial temperature of the ice, assuming no heat loss to the surroundings.

  b) The water, at freezing in (a), is then heated using a further 100 J. What is its final temperature?

G4.3   Calculate the ratio between the energy needed to vaporize a certain quantity of water, and the energy needed to heat that same quantity of water from the freezing to boiling point (without boiling it).

G4.4   2.25 kg of ice, initially at −40 °C, is heated using a 3.2 kW heater without loss to the surroundings. How much time elapses before

  a) the ice reaches melting temperature?

  b) the ice has all melted?

  c) the water reaches boiling point?

  d) the water has all vaporized?

G4.5   0.35 kg of ice at −15 °C is lowered into an insulated beaker of 0.61 kg of water at 59 °C.

  a) What is the temperature after equilibrium has been reached?

  b) What is the minimum mass of water at 59 °C which could be added to achieve a final temperature of 0.0 °C?

  c) What is the maximum mass of water at 59 °C which could be added to achieve a final temperature of 0.0 °C?

Chapter H

Fields

$^8/_{10}$ **H1   Uniform Electric Fields**

In these questions ignore the effects of non-electrical forces.

H1.1   What is the magnitude of the force if a $+6.0 \times 10^{-9}$ C charge is put in a 50 000 N C^{-1} field?

H1.2   What magnitude of field do you need to cause a -1.6×10^{-19} C electron to experience a 2.00 N magnitude force?

H1.3   How strong a field do you need to cause a $+3.2 \times 10^{-19}$ C alpha particle to experience a 2.00 N force?

H1.4   What is the strength of the electric field between two metal sheets held 5.0 cm apart, if one is connected to -500 V, and the other connected to $+2000$ V?

H1.5   What is the field strength needed to cause a spark in air, if 240 V can only jump a distance of 8.0×10^{-5} m?

H1.6   An oil drop with a charge of $+1.2 \times 10^{-15}$ C is held between two horizontal metal plates which are 3.0 mm apart.

   a) If one plate is earthed (i.e. it is at 0 V) and the other is at $+600$ V, what is the force on the oil drop?

   b) If the drop experiences an upwards electrical force, which plate is connected to $+600$ V? The top one or the bottom one?

H1.7   In an accelerator, you want to accelerate electrons (charge $= -1.6 \times 10^{-19}$ C) with a force of 8.0×10^{-12} N as they pass from one metal plate to another. The plates are 2.0 cm apart.

   a) If one plate is earthed (i.e. it is at 0 V) what will be the magnitude of the voltage on the other plate?

   b) If the electrons start on the earthed plate, is the other plate connected to a positive or negative voltage?

H1.8   What is the force on a $+6.0 \times 10^{-13}$ C charge between two metal surfaces if the surfaces are 5.0×10^{-5} m apart, and the potential difference across the plates is 9.0 V?

H2 Electric Field near Point Charges

H2.1 Calculate the magnitude of the force of attraction on a +1.0 nC charge placed 1.0 m away from a −1.0 nC charge.

H2.2 Calculate the electric field strength 1.0 mm away from a +1.0 pC charge.

H2.3 Calculate the magnitude of the repulsive force between two electrons separated by 10^{-10} m.

H2.4 Calculate the magnitude of the attractive force between a proton and an electron separated by 5.0×10^{-11} m.

H2.5 Two +1.0 nC charges are placed 1.0 mm apart. Calculate the electric field strength at the point half way between the charges.

H2.6 1 in 10^{20} of my electrons are removed and given to my wife. This leaves me with a charge of +0.31 C, while my wife has a charge of −0.31 C. Calculate the distance between us when the force of attraction is 2.0×10^5 N (the weight of a bus).

H2.7 I have two mystery charges. When placed 10^{-7} m apart, they experience a repulsive force of 5.0×10^{-4} N. What will be the force between them when they are 4.0×10^{-7} m apart?

H2.8 The electric field 1.0 cm away from a small, strongly charged object is 4.5×10^8 N C^{-1}. What is the charge on the object?

H2.9 At one time, people thought that the electrons in an atom 'orbited' the nucleus like the planets orbiting the Sun. Given that the force needed to keep an electron in its orbit around a proton is 9.2×10^{-8} N, work out the radius of the orbit.

H2.10 Two charges are stuck to a metre stick: a +1.0 pC charge at the 0 cm mark, and a −1.0 pC charge at the 10 cm mark. What is the strength of the electric field at the 20 cm mark? Assume that the wooden metre ruler is strong enough to hold the charges in place, but does not affect the electric field.

H3   Speed of Electron in an Electric Field

For electrons moving at a speed greater than 10% of the speed of light, you should only claim that your answer is approximate (unless you have used relativistic equations). If you reckon that the electron is travelling at a speed greater than 80% of the speed of light, you should decline to give your answer unless using relativity.

H3.1 Convert 7.2 eV into joules.

H3.2 Convert 3.0×10^{-11} J into electronvolts.

H3.3 How fast is an electron going if it has been accelerated from rest by a potential difference of. . .

    a) 50 V

    b) 20 kV

    c) 1.5 GV

H3.4 What accelerating potential is needed to produce electrons with a speed of. . .

    a) 7000 m s^{-1}

    b) 4.0×10^7 m s^{-1}

H3.5 How fast is an alpha particle going if it accelerated by a 1.5 MV potential? Assume that the alpha particle has twice the charge and four times the mass of a proton.

H3.6 To trigger a particular nuclear reaction, a deuterium nucleus (same charge as the proton, but twice the mass) needs to have a kinetic energy of 4.0×10^{-13} J. What accelerating voltage is needed?

H3.7 In order to produce protons with a kinetic energy of 5.0 MeV, what accelerating voltage is needed?

H4   Force on a Conductor in a Magnetic Field

Ignore the Earth's magnetic field unless specifically asked. Assume the horizontal component of the Earth's field points North, and the vertical down.

The horizontal component of the Earth's magnetic field in Britain is 6.91×10^{-5} T, while the vertical component is 1.55×10^{-4} T.

H4.1   Complete the questions in the table:

Length	Current	Field	Field–current angle	Force
0.30 m	10 A	5.0 T	90°	(a)
2.0 cm	200 A	0.040 T	30°	(b)
56 m	20 mA	(c)	90°	3.5 N
2.0 m	(d)	7.0×10^{-5} T	90°	0.032 N

H4.2   Calculate the force on 3.0 mm of wire carrying a 4.0 A current in a 0.020 T field, if the current is perpendicular to the field.

H4.3   Calculate the force on 3.0 mm of wire carrying an 8.0 A current in a 0.0040 T field, if the current is parallel to the field.

H4.4   Calculate the current needed in a wire if you wish it to levitate in a 0.50 T field. Assume that the wire has a weight of 0.14 N, and 3.0 cm of it is inside the magnetic field region.

H4.5   There is a bad electrical fault in a house. A 6.0 m wire running North-South carries a current of 6000 A for a short time.

   a) Calculate the magnitude of the force on it due to the Earth's magnetic field.

   b) At an instant when the current in the wire is running from North to South, what is the direction of the force?

H4.6   A rail gun consists of two metal rails 8.0 cm apart in a magnetic field with a projectile (shell) placed to bridge the rails. A large current is passed from one rail to the other through the projectile, which then experiences a force and shoots off the end of the rails. A 100 N force needs to act on the projectile. Assuming permanent magnets provide a 0.16 T magnetic field, how much current needs to pass through the projectile?

H5   Force on Particle in a Magnetic Field

H5.1   Complete the questions in the table:

Charge /C	Speed /m s^{-1}	Angle between velocity & B-field /°	Magnetic Flux Density /T	Force /N
6.0×10^{-9}	0.45	90	1.3	(a)
2.0×10^{-12}	31	30	0.00056	(b)
2.0×10^{-17}	(c)	90	8.4	3.2×10^{-15}

H5.2   Calculate the force on an electron going at 3.5×10^7 m s^{-1} in a 3.4 mT magnetic field:

  a) If the electron is travelling perpendicular to the magnetic field.

  b) If the electron is travelling parallel to the magnetic field.

H5.3   An electron is travelling at right angles to a magnetic field, and at right angles to an electric field such that the electric and magnetic forces cancel out. If the magnetic flux density is 0.043 T and the electric field is 330 kV m^{-1}, how fast is the electron going?

H6 Circular Paths of Particles in Magnetic Fields

H6.1 Complete the questions in the table:

B /T	q /C	v /m s^{-1}	m /kg	r /m
0.63	1.6×10^{-19}	3.0×10^{7}	9.1×10^{-31}	(a)
0.63	1.6×10^{-19}	6.4×10^{6}	1.7×10^{-27}	(b)
2.30	3.2×10^{-19}	8.8×10^{7}	(c)	0.80
0.0045	1.6×10^{-19}	(d)	9.1×10^{-31}	0.12

H6.2 In a demonstration, electrons with 200 eV of kinetic energy are going round in a 12 cm **diameter** circle. Calculate the magnetic flux density.

H6.3 In a demonstration, electrons are going round in a 12 cm diameter helix with the beam at 70° to the 0.0032 T magnetic field. Calculate the speed of the electrons.

H6.4   a) Work out the momentum of a muon (same charge as an electron, but mass = 207 × electron mass) taking a curved path with a 90 cm radius perpendicular to a 0.0076 T magnetic field.

   b) Work out the momentum of an electron which would take the same path in the same field.

H7 Magnetic Flux and Faraday's Law

H7.1 Complete the questions in the table:

Magnetic Flux Density /T	Area of Coil	Angle between plane of coil and magnetic field lines /°	Num-ber of turns	Magnetic flux linkage /Wb turns
2.0	2.0 m × 1.0 m	90	40	(a)
0.00232	5.0 cm × 5.0 cm	60	2400	(b)

H7.2 Calculate the magnetic flux linkage if a 3.0 cm × 2.0 cm rectangular coil of 200 turns is in a 0.75 T magnetic field, with the field at right angles to the plane of the coil.

H7.3 Calculate the magnetic flux linkage if a 2400 turn coil measuring 3.0 cm × 3.0 cm lies within a 0.25 T magnetic field, with the field lines making an angle of 30° to the plane of the coil.

H7.4 Assume field lines are perpendicular to the plane of a 400 turn coil of area 3.0×10^{-4} m^2.

   a) Calculate the rate of change in the magnetic flux linkage when the magnetic field is reduced from 0.20 T to zero in 0.40 s.

   b) What is the voltage induced across the coil?

H7.5 Complete the questions in the table:

Initial flux linkage /Wb turns	Final flux linkage /Wb turns	Time taken for flux to change /s	Voltage induced /V
30	60	0.20	(a)
200	0	(b)	400

H7.6 A single turn coil of 10 cm × 5.0 cm sits, stationary, in a 21000 T magnetic field, at right angles to the plane of the coil.

   a) What is the voltage induced across the ends of the wire?

   b) The coil is made of extensible wire and is stretched steadily to 10 cm × 10 cm over 0.020 s. Calculate the voltage induced across the ends of the wire.

   c) What would the induced voltage be if the magnetic field were parallel to the sides of the coil which were originally 5.0 cm long?

H7.7 A bicycle wheel with only one spoke has a magnetic flux of 1.95×10^{-5} Wb passing through it. If the wheel goes round 6 times in one second, what voltage will be induced between the hub and the rim?

Something to think about – would the answer to question H7.7 change if there were twenty spokes?

H8 Transformers

All transformers are perfectly efficient unless you are told otherwise.
Complete the questions in the table:

	Turns on primary	Turns on secondary	Primary voltage /V	Secondary voltage /V	Step up or step down
H8.1	2400	(a)	230	12	(b)
H8.2	1200	60	(a)	4.5	(b)
H8.3	(a)	500	275000	11000	(b)
H8.4	20000	(a)	11000	230	(b)
H8.5	20	1000	15000	(a)	(b)

H8.6 You have a 230 V supply capable of delivering 13 A. An experiment requires a current of 200 A that you take from a step-down transformer. If there are 1200 turns on the primary, how many turns should there be on the secondary?

H8.7 What is the secondary voltage across a transformer when the primary has been attached to a 12 V battery? The primary has 2000 turns, and the secondary has 3000 turns.

H8.8 A shaver socket is powered from the 230 V mains via a transformer. The primary and secondary have 800 turns. What is the output voltage?

H8.9 Calculate the current in the load fed by the secondary of a 90% efficient step down transformer where the primary has 50× as many turns as the secondary, and where the primary current is 5.0 A?

H8.10 A loudspeaker is powered from the secondary of a transformer with a turns ratio of 10:1 (i.e. primary to secondary). The loudspeaker has an 8.0 Ω resistance, which means that when the secondary voltage is 10 V, the current flowing in the secondary is 1.25 A. What resistance does the speaker appear to have when measured on the primary side of the transformer (i.e. take resistance as primary voltage divided by primary current)? Neglect the resistance of the transformer itself.

H9   Energies and potentials of charges in electric fields

All answers must be given with the correct sign. Questions H9.1 to H9.3 concern the region between two large, horizontal metal plates 2.00 mm apart that are connected to a 1.60 kV power supply. The negative terminal is earthed and connected to the bottom plate. Ignore complications caused by the edges of the plates.

H9.1   Calculate the potential of a point:

    a) 1.00 mm above the bottom plate.

    b) 0.75 mm above the bottom plate.

H9.2   How far above the bottom plate would a point need to be if its potential was 1.35 kV?

H9.3   Calculate the electrostatic potential energy of

    a) a proton 1.85 mm below the top plate.

    b) an electron 0.32 mm above the bottom plate.

H9.4   Calculate the electrostatic potential of a point 0.92 m away from a very small +24 nC charge.

H9.5   A metal sphere with a radius of 7.4 cm is at a potential of 1.8 MV.

    a) Calculate the charge stored on the sphere.

    b) Calculate the potential of a point 13.6 cm from the sphere's centre.

H9.6   Calculate the electrostatic potential energy when a proton is 0.43 nm from an electron.

H9.7   An alpha particle (charge of 3.2×10^{-19} C and mass 6.7×10^{-27} kg) is fired directly towards a gold nucleus which has a charge of 1.26×10^{-17} C.

    a) Taking a speed of 1.8×10^{7} m/s, and assuming negligible recoil of the gold nucleus, calculate the distance of closest approach.

    b) How fast would the alpha particle have to be fired in order to have a closest approach distance of 8.0×10^{-15} m?

H9.8   Two charges are stuck to a metre stick: a +1.0 pC charge is stuck to the 0.0 cm mark, and a −1.0 pC charge is stuck to the 10 cm mark.

    a) Calculate the electrostatic potential at the 20.0 cm mark.

    b) Find the electrostatic potential at the 5.0 cm mark.

Chapter I

Capacitors

I1 Charge and Energy stored on a Capacitor

Complete the questions in the table:

	Capacitance	Voltage /V	Charge /C	Energy /J
I1.1	100 μF	6.0	(a)	(b)
I1.2	(a)	12.0	(b)	0.0010
I1.3	(a)	240	1.6×10^{-4}	(b)
I1.4	10 nF	(a)	1.6×10^{-4}	(b)

I1.5   Calculate the capacitance of a capacitor needed in a back up power supply if it needs to store 0.24 J of electrical energy when connected to a 12 V power supply.

I1.6   When a metal strip is rubbed on a 5000 V terminal, it gains 6.0 nC of charge. Calculate the effective capacitance of the strip.

I1.7   A 2200 μF capacitor needs to be able to supply an average current of 2.0 mA for five minutes. Calculate the charge needed, and therefore the operating voltage which has to be employed.

I1.8   A mystery capacitor can store 3.0 J of energy when connected to a 10 V supply. How much energy can it store when connected to a 5.0 V supply?

I2 Capacitor Networks

I2.1   Calculate the capacitance of each of the following combinations:

   a) A 3.0 μF capacitor connected in parallel with a 2.0 μF capacitor.

   b) A 3.0 μF capacitor connected in series with a 2.0 μF capacitor.

   c) A 6.0 μF capacitor is connected in parallel with a 4.0 μF capacitor. The combination is then connected in series with a 20 μF capacitor.

   d) A 220 nF capacitor is connected in series with a 440 nF capacitor. The combination is connected in parallel with a 1.0 μF capacitor.

   e) A 1.0 nF, 2.0 nF and 3.0 nF capacitor, all connected in parallel.

   f) A 1.0 nF, 2.0 nF and 3.0 nF capacitor, all connected in series.

I2.2   A 200 μF capacitor is in series with a 2200 μF capacitor and they are charged until the 200 μF capacitor stores 30 μC. What is the charge on the other capacitor?

I2.3   A 200 μF capacitor is in series with a 2200 μF capacitor. The capacitors are charged until the 200 μF capacitance has a voltage of 12 V across it. What is the voltage across the 2200 μF capacitor?

I2.4   A 470 μF capacitor is charged using a 10 V battery. It is then disconnected, and connected to an uncharged 220 μF capacitor. Calculate the voltage across the capacitors once the current has stopped flowing. (Hint: capacitors are effectively in parallel, and total charge has not changed.)

I2.5   A 6.0 nF capacitor is in parallel with a 10 nF capacitor. The voltage across the 6.0 nF capacitor is 36 V. What is the voltage across the other capacitor?

I3 Discharge of a Capacitor

Complete the questions in the table:

	Capacitance	Resistance	Time constant	Halving time
I3.1	100 μF	200 kΩ	(a)	(b)
I3.2	2200 μF	(a)	45 s	(b)
I3.3	(a)	330 Ω	(b)	0.10 s
I3.4	10 μF	(a)	3.0 minutes	(b)

I3.5   Draw the circuit diagram for a circuit which could discharge a capacitor through a fixed resistor while measuring the discharge current and voltage across the capacitor.

I3.6   A 2200 μF capacitor is charged with a 12 V battery. It is then discharged through a 10 kΩ resistor.

   a) What is the initial discharge current?

   b) Calculate how long the capacitor would take to discharge if the initial rate of discharge were maintained.

   c) What will the voltage be across the capacitor after 22 s?

   d) What will the current be when the voltage across the capacitor has halved?

   e) How much time will it take before the capacitor has a voltage of 3.0 V across it?

I3.7   A 5.0 μF capacitor is charged with a 20 V supply. It is then discharged through a 10 kΩ resistor.

   a) Calculate the time taken for the voltage across the capacitor to halve.

   b) Calculate the voltage across the capacitor two time constants after the discharging starts.

   c) Calculate the charge on the capacitor after one time constant.

      d) Calculate the current flowing in the circuit 0.20 ms after the discharging starts.

I3.8   If you want to make a timing circuit where a capacitor's voltage reduces from 12 V to 4.0 V over 3.0 minutes using a 1000 μF capacitor, what value of resistance do you need?

I3.9   A 500 μF capacitor is initially uncharged. It is connected to a 12 V battery in series with a 20 kΩ resistor. Work out the voltage across the resistor after 8.0 s.

Chapter J

Nuclear Physics

J1   Nuclear Equations

Complete the nuclear equations. Don't forget the neutrino / antineutrino if it is a beta decay!

J1.1 $^{241}_{95}\text{Am} \rightarrow \,^{?}_{?}\text{Np} + ?$ Alpha decay

J1.2 $^{3}_{1}\text{H} \rightarrow \,^{?}_{?}\text{He} + ?$ Beta- decay

J1.3   (a) $^{14}_{6}\text{C} \rightarrow \,^{?}_{?}\text{N} + ?$ Beta- decay   (b) $^{11}_{6}\text{C} \rightarrow \,^{?}_{?}\text{B} + ?$ Beta+ decay

J1.4 $^{60}_{27}\text{Co}^{+} \rightarrow \,^{?}_{?}\text{Co} + ?$ Gamma decay

J1.5 $^{3}_{1}\text{H} + \,^{2}_{1}\text{H} \rightarrow \,^{?}_{?}\text{He} + \,^{?}_{?}\text{n}$ Nuclear Fusion

J1.6 $^{90}_{38}\text{Sr} \rightarrow \,^{?}_{?}\text{Y} + ?$ Beta- decay

J1.7 $^{238}_{92}\text{U} \rightarrow \,^{?}_{?}\text{Th} + ?$ Alpha decay

J1.8 $^{235}_{92}\text{U} + \,^{1}_{0}\text{n} \rightarrow \,^{147}_{57}\text{La} + \,^{87}_{?}\text{Br} + ?^{?}_{?}\text{n}$ Nuclear Fission

J1.9 $^{23}_{13}\text{Al} \rightarrow \,^{?}_{?}\text{Mg} + ?$ Beta+ decay

And as a bonus round, the unattached neutron is unstable too.

J1.10 $^{?}_{?}\text{n} \rightarrow \quad ? + ? + ?$ Beta- decay

J2   Activity and Decay

A 'mole' of nuclei contains 6.02×10^{23} nuclei. The mass of one mole of nuclei (the 'molar mass') is approximately equal to 0.001 kg × the mass number of the nucleus. Use this approximation wherever a question does not give the molar mass explicitly.

Complete the questions in the tables:

	Half life	Decay constant /s^{-1}	Half life	Decay constant /s^{-1}
J2.1	53 s	(a)	12 years	(b)
J2.2	(a)	3.2×10^{-10}	(b)	1.2×10^{-4}

	Decay constant /s^{-1}	Activity /Bq	Number of nuclei	Mass of sample /kg	Molar mass /kg
J2.3	0 (isotope stable)	(a)	(b)	2.4×10^{-4}	0.012
J2.4	0.0138	230	(a)	(b)	0.085
J2.5	3.42×10^{-11}	5600	(a)	(b)	0.239
J2.6	1.83×10^{-9}	(a)	(b)	3.0×10^{-5}	0.003

J2.7   a) How many nuclei are there in 5.0 mg of ^{14}C?

b) What is the activity of the sample, if the half life is 5700 years?

J2.8   a) ^{238}U has a half life of 4.47×10^9 years. How many ^{238}U nuclei are needed for an activity of 5000 Bq?

b) What is the mass of the ^{238}U sample?

J2.9   Long half lives are measured using the principle of activity. If 3.0 mg of ^{239}Pu has an activity of 6.9×10^6 Bq, calculate the half life of ^{239}Pu.

J2.10   A 'radioactive battery' for a long range space probe uses a radioisotope with a decay constant of 4.4×10^{-12} s^{-1}, and a molar mass of 0.236 kg. Each time one nucleus decays, 2.5×10^{-12} J of electrical energy is output by the generator. Calculate the mass of the radioactive sample if the spacecraft requires 200 W of electricity.

J3 Nuclear Decay with Time

Complete the questions in the table:

	Initial number of unstable nuclei	Initial Activity /Bq	Half-life	Decay constant /s^{-1}	Number of unstable nuclei left after 6 hours	Activity after 6 hours /Bq
J3.1	(a)	3000	6.0 h	(b)	(c)	(d)
J3.2	2.0×10^{21}	(a)	3.0 h	(b)	(c)	(d)
J3.3	(a)	23000	(b)	1.28×10^{-4}	(c)	(d)
J3.4	(a)	700	(b)	(c)	(d)	200

J3.5 Tritium has a half life of about 12 years. If you put 3.0 μg of tritium into a luminous sign, how much will still be there 50 years later?

J3.6 If a substance has a half life of 100 s, how long do you have to wait for 25% of the nuclei to decay?

J3.7 A substance has a half life of 100 s, and starts with 10^{20} unstable nuclei.

    a) Calculate the initial activity, and from this work out the time taken for all of the nuclei to decay if the activity did not decrease with time.

    b) Calculate what fraction of the nuclei remain after the time calculated above.

J3.8 Carbon-14 has a half life of about 5700 years. What fraction of the original amount of carbon-14 would you expect to find in the timbers of a boat built 8000 years ago?

J3.9 Uranium-238 has a half life of 4.47×10^{9} years and decays to thorium-234. The thorium decays (by a series of further nuclear processes which are relatively brief) to lead. Assuming that a rock was originally entirely uranium, and that at present, 1.5% of the nuclei are now lead, calculate the age of the rock.

J4   Energy in Nuclear Reactions

Mass defects, binding energies or energy yields in nuclear reactions require high precision data as calculations involve subtracting two very similar numbers. Use only data here (& on page iv), to **all** significant figures given: **take** $c = 2.998 \times 10^8$ m s^{-1}, **and the electronic charge as** 1.602×10^{-19} C.

J4.1   Calculate the mass defect of $^{56}_{26}$Fe in kilograms. The ^{56}Fe **nucleus** has a mass of 55.92068 u.

J4.2   Calculate the mass defect of $^{12}_{6}$C in kilograms. The ^{12}C **atom** has a mass of 12.00000 u.

J4.3   Calculate the binding energy of ^{56}Fe in MeV.

J4.4   Calculate the binding energy per nucleon of ^{12}C in MeV.

J4.5   One nuclear fusion reaction is $^2_1$H + $^3_1$H $\rightarrow$ $^4_2$He + $^1_0$n. The masses of the **nuclei** are given below:

Deuterium (^{2}H) mass	2.013 55 u
Tritium (^{3}H) mass	3.015 50 u
Helium (^{4}He) mass	4.001 51 u

Calculate the energy released by this reaction in MeV (it appears as the kinetic energy of the reaction products).

J4.6   One nuclear fission reaction is $^{235}_{92}$U + $^1_0$n $\rightarrow$ $^{147}_{57}$La + $^{87}_{35}$Br + 2^1_0n. The masses of the **atoms** are given in the table below. Calculate the energy released by this reaction in MeV.

^{235}U	$3.903\,00 \times 10^{-25}$ kg
^{147}La	$2.439\,81 \times 10^{-25}$ kg
^{87}Br	$1.443\,35 \times 10^{-25}$ kg

J4.7   a) Using the J4.5 data of nuclear masses, calculate the binding energy per nucleon in deuterium.

b) Calculate the energy released in the fusion reaction of J4.5, using the result in (a) and: Binding energy per nucleon of tritium is 2.8273 MeV, and of helium-4 is 7.0739 MeV.

J4.8   a) Using the J4.6 data table of atomic masses, calculate the binding energy per nucleon in ^{235}U in MeV.

b) Calculate the energy released in the fission reaction of J4.6, now using the result in (a) and the following data: The binding energy per nucleon for ^{147}La is 8.2227 MeV, and for ^{87}Br is 8.6055 MeV.

Chapter K

Modelling the Universe

K1   Red Shift and Hubble's Law

Complete the questions in the table:

	Emitted Wavelength /nm	Received Wavelength /nm	$\Delta\lambda$ /nm	Red shift as fraction of emitted λ = recession velocity as fraction of c	Recession velocity /m s^{-1}
K1.1	500	550	(a)	(b)	(c)
K1.2	500	(a)	(b)	(c)	3.0×10^6
K1.3	500	480	(a)	(b)	(c)
K1.4	(a)	663	(b)	(c)	-2.1×10^7
K1.5	300	1.1×10^6	(a)	(b)	(c)

K1.6   A specific wavelength called the 'calcium K-line' is measured on Earth in laboratories (using a stationary calcium plasma) as 393.4 nm. A galaxy is moving away from our own at a speed of 100 km s^{-1}. It is observed using telescopes on Earth, and the apparent wavelength of the calcium K-line is measured. What is the value of the wavelength as measured through the telescope?

   a) What is the value of the wavelength as measured through the telescope?

   b) Assuming a value of the Hubble constant of 70 km s^{-1} Mpc^{-1}, calculate the distance of the galaxy from the Earth.

K1.7   The Ursa Major cluster of galaxies is about 200 Mpc from Earth.

a) How far away is this in metres?

b) Convert the value for the Hubble constant in question K1.6(b) into S.I. units (s^{-1}).

c) Estimate the recession velocity of the Ursa Major cluster in m s^{-1}.

NB - Question K1.5(c) is not an error. This is data for a wavelength within the cosmic microwave background radiation. Hopefully this will enable you to see that there is more to cosmic red-shift than the Doppler effect could ever explain! It turns out that if the Universe (the fabric of space time) expands while light is in transit, the wavelength of the light gets stretched by the factor of expansion. So you don't need a superlumic speed of recession to account for the very large red-shift in question K1.5(c) – you just need the Universe to have got nearly 4000 times bigger in the fourteen billion years or so in which the background radiation has been out there. . .

K2 Exponential Extrapolation

It is advisable to have completed section J3 before beginning this one.

K2.1 If 45% of the unstable nuclei in a sample take 50 s to decay, calculate the decay constant in s^{-1}.

K2.2 If 70% of the light falling on a 5.0 mm thick block of coloured material emerges from the other side, calculate the attenuation coefficient of the material in mm^{-1}.

K2.3 After a period of 3.0 minutes, only 20% of the original charge remains on a capacitor. Calculate the time constant RC of the circuit.

K2.4 In a stage light, 8.0 W of light pass into a 0.70 mm thick filter, of which 6.5 W is absorbed. Calculate the attenuation coefficient of the material.

K2.5 A sample has an initial activity of 3300 Bq. After 15 minutes, the activity is 1230 Bq. What will the activity be after a further 15 minutes?

K2.6 The voltage across a capacitor is 11.5 V. One hour later, it is 7.2 V. What will the voltage be 3.0 hours after the original measurement?

K2.7 It is said to be safe to view the Sun through a filter if it only lets 10^{-5} of the light through. Suppose you have some material which lets 2.0% of the light through. How many sheets do you need to put together back-to-back before you can safely look through it at the Sun? NB - Never make your own filter for viewing the Sun in this way - most filters bleach with very high intensities and aren't designed with eye protection in mind, so the quality is not good enough for a device which is to prevent blindness.

K2.8 The attenuation coefficient for a particular beta decay is 2.4 mm^{-1} through aluminium. What thickness of aluminium is needed to reduce an initial count rate of 5.0×10^5 s^{-1} to background levels of 5.0 Bq?

K2.9 The thickness of lead needed to stop half of the neutrinos in a beam is about 3000 light years (which you may take as 3.0×10^{19} m). Calculate the fraction of neutrinos which would be stopped by 100 m of water assuming that the attenuation co-efficients for water and lead are about the same (which they're not).

Optional, but related, and useful:

K2.10 You start with a credit card debt of £150. For each month in which you don't pay it off, the debt increases by 3.0%. Assuming you pay nothing for 3.0 years, and then want to settle the debt in one go, how much would you have to pay?

Chapter L

Fact Sheets

L1 Mass Spectrometers

L1.1   What does a mass spectrometer actually measure?

L1.2   a) What kind of particles move through the spectrometer?

b) Give one method of producing the particles.

L1.3   a) The first region of the spectrometer contains a field - is it electric or magnetic? Is it aligned parallel or perpendicular to the beam?

b) What will happen to a heavier particle compared to a light one of the same charge?

L1.4   The next region sometimes includes a device with electric and magnetic fields set up at right angles. What does this do?

L1.5   a) The final region of the spectrometer contains a field - is it electric or magnetic? Is it aligned parallel or perpendicular to the beam?

b) What is the shape of the track of the particles in this part of the spectrometer?

c) What will happen to a heavier particle compared to a light one of the same charge?

L1.6   In Time of Flight mass spectrometry the surface of the material is struck by a _____ beam, and batches of molecules are _____ and _____ towards a detector. The mass can be worked out from the time it takes the molecules to reach the detector. Fill in the blanks.

L1.7   Give an advantage of time of flight mass spectrometry over the conventional kind.

L1.8   Give a practical application of mass spectrometry.

L2 Fundamental Particles and Interactions $^9/_{12}$

L2.1 State the category of fundamental particles which contains the electron.

L2.2 Give the quark content of the neutron.

L2.3 Which kind of radioactive decay gives out an antineutrino?

L2.4 What is the

    a) charm number of a particle containing quarks *udc*?

    b) strangeness number of a particle containing quarks *uds*?

L2.5 What is a baryon?

L2.6 Under what circumstances is

    a) baryon number conserved?

    b) strangeness conserved?

L2.7 Which particle mediates the electromagnetic force?

L2.8 How does the mass of an antiparticle compare with the equivalent ordinary particle?

L2.9 What is the baryon number of a top quark?

L2.10 What is the lepton number of the electron antineutrino?

L3   Nuclear Reactors

L3.1   What is a control rod typically made of?

L3.2   What is the function of a moderator?

L3.3   If a normal reactor had no moderator the reaction would stop. Why?

L3.4   Give the term which describes the fact that one fission gives out neutrons which cause fission which makes more neutrons which cause more fissions...

L3.5   What is a thermal neutron?

L3.6   Which material is used as the coolant in most nuclear reactors?

L3.7   Which isotope is typically used as fuel in a nuclear reactor?

L3.8   Most nuclear reactors have two cooling loops, one heating the other. Why?

L3.9   Nuclear waste is very radioactive. Why?

L3.10   a) Why is a high temperature needed to sustain a fusion reaction?

      b) Give one other condition needed for a sustained fusion reaction in a lab.

      c) In the lab, how are the conditions needed for fusion produced?

L4   X-Rays

L4.1   a) Give a typical energy of an X-ray photon in electronvolts (eV).

      b) What accelerating voltage would be required to produce an electron of the same kinetic energy?

      c) Give the name of the process by which electrons are freed from the heated cathode.

L4.2   a) Sketch the X-ray spectrum produced by a typical X-ray machine (give a graph of intensity vs frequency) with no filtering.

      b) In a different colour, add to your sketch the spectrum expected from the same machine if the accelerating voltage were doubled.

L4.3   What is the predominant process occurring when 100 keV electrons interact with atoms?

L4.4   How is the **Compton effect** different from the **photoelectric effect**? After all, they both free an electron from an atom...

L4.5   Often, a thin metal filter is placed between the X-ray machine and the patient. Why?

L4.6   High energy X-rays can be used for radiotherapy (cancer treatment), as they have the same frequencies as gamma rays.

      a) Why are they still called X-rays even though they have the same frequency as a gamma ray?

      b) Why do medical professionals often prefer to use these high energy X-rays rather than 'real' gamma rays?

L4.7   Give three differences between a conventional X-ray and a CAT scan.

L4.8   Draw a diagram to show how the use of a lead grid can improve the quality of an X-ray image.

L4.9   Name a contrast medium used with conventional X-rays to assist in the imaging of tissue which would not usually show up on an X-ray.

L4.10   a) Define **intensity**, and give its unit.

      b) Define the **attenuation coefficient** μ, and give its unit.

      c) Give a typical value for μ for bone.

L5   Ultrasound

L5.1   Give a typical frequency of ultrasound as used in medical imaging.

L5.2   Give the name for the phenomenon which allows certain crystals to produce ultrasound when an alternating voltage is applied to them.

L5.3   a) In one word: what do ultrasound waves reflect off?

b) The strength of the reflection is determined by...

c) Why can't you use ultrasound to image the lungs?

L5.4   If a single beam of ultrasound waves is used, an A-scan is produced. Sketch what an A-scan trace might look like. Make sure that you label the axes with what is being plotted.

L5.5   A B-scan gives a 2-dimensional image. How does the scan head for a B-scanner need to be different to that used only for A-scans?

L5.6   Why is the patient smeared in gel before an ultrasound scan is conducted?

L5.7   Why is ultrasound the wave of choice for pre-natal scanning?

L5.8   Give one **therapeutic** use of ultrasound (i.e. a medical use where the ultrasound treats a condition rather than taking pictures of it).

L5.9   If the reflection off the back of a foetus's head takes 0.092 ms longer to reach the receiver than the reflection off the front of the head, calculate the size of the foetus's head. Assume that the speed of ultrasound is 1400 m s^{-1}.

L5.10 What can **Doppler ultrasound** be used for?

L6 MRI and PET scanning

L6.1 a) State an isotope which can be injected into a patient in advance of a PET scan.

b) State which fundamental particle is detected in the 'camera' during a PET scan.

L6.2 How can the computer tell the difference between background counts and genuine 'signal' counts whilst a PET image is being taken?

L6.3 a) On a PET scan, a bright area of the scan implies what?

b) Give one condition which can be identified or diagnosed using a PET scan.

L6.4 How do you make the radioisotope typically used in PET scanning?

L6.5 a) State the material typically used to produce flashes of visible light.

b) State the device used to pick up the minute amount of visible light produced in this detector and convert it to a measurable electrical signal.

L6.6 a) State a typical value for the magnetic flux density inside an MRI scanner.

b) Give the frequency which corresponds to this magnetic field. One significant figure is sufficient.

L6.7 A hydrogen nucleus, with its lone proton has a magnetic moment. What happens to the proton's magnetic field when it is put in a strong magnet, such as during an MRI scan?

L6.8 What is the name of the frequency of radiation which will resonate with a given atom, causing its magnetic orientation to flip?

L6.9 a) When an atom 'relaxes' and gives out waves, how can the scanner/computer work out where the atom was?

b) How does the scanner/computer work out the type of tissue/material at a place where relaxation has just occurred?

L6.10 a) Give one reason or circumstance when MRI scanning would be preferred to a PET scan.

b) Give one reason or circumstance when PET scanning would be preferred to an MRI scan.

L7   Stars

L7.1   Which elements were produced in the Big Bang?

L7.2   a) What is the name of a collapsing mass of dust and gas which has not yet reached the temperatures and pressures needed for ignition?

b) What is the temperature in the centre of the Sun?

L7.3   What is the nuclear reaction which powers stars?

L7.4   Which two forces govern the progress of a star, and are in equilibrium when the star is in a steady phase?

L7.5   a) What is the term for a stable star (such as our Sun) in the 'adult' phase of its existence?

b) Other than yellow stars such as the Sun, give another type of star which is also in the 'adult' phase of its existence.

c) Give the difference in nuclear process between a regular 'adult' star and a red giant/supergiant?

L7.6   Give the time period for which a star will typically remain as a red giant/supergiant.

L7.7   Red giants and red supergiants both end in explosions. But the explosions are different. Give the differences as listed below

a) What is the last element formed in bulk before the explosion?

b) What is the name of the explosion?

c) What is formed in the core of the giant as a result of the explosion?

d) In the case of a red giant - what is the name given to the space object formed from the material ejected during the explosion.

L7.8   On Earth, there is quite a lot of uranium. Where was it made?

L7.9   How heavy does the core of a red supergiant star have to be in order to form a black hole?

L7.10 What is a pulsar? (Make sure you give the reason for the pulsing.)

L8   The History of the Universe

L8.1   Roughly how long ago was the Big Bang?

L8.2   Which chemical elements were formed as a result of the Big Bang? Give the percentage abundance (by mass) of the main two.

L8.3   Approximately how much time elapsed after Creation before protons and neutrons were formed from the 'quark soup'?

L8.4   A very important event occurred when the atoms formed from the loose nuclei and electrons (plasma).

   a) How much time passed between Creation and the formation of the atoms?

   b) What was the approximate temperature of the Universe at that time?

   c) Which band of electromagnetic radiation was brightest in the Universe at that time?

   d) Why is it impossible to 'see' events earlier than the atomic formation by looking at very distant objects using optical telescopes?

   e) The radiation emitted during this process is still out there. What do we now call it?

   f) How uniform is this radiation?

L8.5   Originally, the four forces were indistinguishable. Give the chronological order in which the forces separated off from the original Unified Force.

L8.6   The average density of the Universe might be less than, equal to or greater than the 'critical density', ρ_0. Assuming that the escape velocity when at distance R from a mass M is given by $v = \sqrt{\frac{2GM}{R}}$, derive an expression for ρ_0 in terms of the Hubble constant H_0 if Newton's Law of Gravitation holds for the Universe at large.

L8.7   In a Universe obeying Newton's Law of Gravitation, what will eventually happen if its density is:

   a) equal to ρ_0?

   b) greater than ρ_0?

L8.8   What is meant by a 'closed' Universe?

L8.9   The Cosmological Principle assumes that the Universe is _____ and _____. Give these words and their meaning in normal language.

L8.10 Olbers' Paradox tells us that the Universe can't be both _____ and _____, given that the night sky is black.

Isaac Physics

Developing problem solving skills.

L. Jardine-Wright
Director, Isaac Physics Project

About the author

Lisa Jardine-Wright is an astrophysicist, and a tutor and director of studies in physics at Churchill College, Cambridge. She lectures first year maths and physics within the Cambridge Natural Sciences degree, and has also directed and developed the educational outreach programmes of the Cavendish Laboratory.

In 2017, the University of Cambridge awarded her the Pilkington Prize for her outstanding and sustained contribution to excellent teaching in Physics and Mathematics both within the University and beyond and in 2019, she was jointly awarded the IOP Lawrence Bragg Medal and Prize, with Prof. Mark Warner, for the foundation and development of the Isaac Physics project. She was elected as the Vice-President for Education of the Institute of Physics in 2020.

Periphyseos Press
Cambridge, UK.

Acknowledgements

In this book we provide a sample of physics problems, selected from thousands that are available at isaacphysics.org. The Isaac Physics project and I would like to thank Cambridge Assessment for their permission to include selections of their previous examination questions on the Isaac platform, an Open Platform for Active Learning (OPAL).

I would like to record grateful thanks to all of the Isaac team past and present, and to the computer scientists that so ably produce the software and infrastructure that makes isaacphysics.org unique.

LJW, 2020

The Isaac Physics Project

The Isaac Physics project is funded by the Department for Education in England and hosted by the Cavendish Laboratory and the Department of Computer Science & Technology of the University of Cambridge.

Isaac Physics: Additional Features and Resources

In additional to this book, Isaac Physics also offers:

- **More books** - How to solve Physics Problems, Pre-Unviersity Maths for Sciences, Mastering Essential Pre-University Physical Chemistry, Mastering Essential GCSE Physics.   https://isaacbooks.org/

- **Master Maths** - Complete the "top 10" questions in pure, mechanics or further maths to practice of all of the topics in each course.
  isaacphysics.org/pages/master_maths

- **Automarking** - create virtual classes and set questions to be automarked by Isaac.
  isaacphysics.org/teacher_features

- **Free events** - workshops and CPD sessions, and online tutorials.
  isaacphysics.org/events

- **Mentoring scheme** - programmes for teachers and students at both GCSE and A level. Work is set weekly and supported by live Q&A every second week with the Isaac team.   isaacphysics.org/pages/isaac_mentor,
  isaacphysics.org/pages/teacher_mentoring

- **Extraordinary questions** - for those who would like more of a challenge.
  isaacphysics.org/extraordinary_problems

- **GCSE resources** - physics skills mastery book, quick quizzes, preparation for A level.
  isaacphysics.org/gcse

Contents

Developing Physics Problem Solving

Isaac Physics - You work it out `isaacphysics.org`

The key concepts and numerical manipulations that you practise in Chapters A — L are the vital step on the road to studying physics, engineering or any STEM course at university. To further advance your thinking as a physicist, you also need to enhance these skills and this analysis by combining multiple concepts and techniques, in order to solve longer and more involved problems. These multi-step, multi-stranded questions can be challenging for experienced physicists, but completion of such problems gives a real sense of achievement and success.

The main mission of Isaac Physics is to provide problem solving questions for students at GCSE (or equivalent) through to those who have finished their A-levels and are preparing for university. We have over 850 questions graded from Level 1 (post-GCSE) through to Level 6 (pre-university and beyond – NOT for the fainthearted).

In the following chapters you will find:

- advice on how to problem-solve,

- sample solutions to two problems from our site (Level 3 Dynamics: Pop-up Toy, and Level 5 Statics: Prism)

- sample questions from mechanics and fields, from our six levels on isaacphysics.org
  Online, we also have the topics: waves, circuits and physical chemistry. They too span all six levels, with questions for you to try.

- next to each topic heading, a url link to a webpage containing this and the other questions from the same level and topic.

Take up the challenge and put your new found skills to the test — but before you do, make sure you pick up a large **pad of paper** and **a pencil** so that *you* can work it out.

A Guide to Solving Physics Problems

isaacphysics.org/solving_problems

Physicists develop highly desirable skills through their extensive experience of problem solving — logic, determination, resilience and mathematical ability to name just a few. You will become an ace problem solver by answering *lots* of questions — these 5 key steps will help you develop a logical, structured method and universal approach.

---

5 key steps to problem solving

- Step 1: Keywords

- Step 2: Diagram

- Step 3: Concepts

- Step 4: Symbols

- Step 5: Dimensions & Numbers

When faced with a new question, we employ a strategy to break the problem down into a series of 5 steps to digest and analyse it. Each step helps us to understand the information given in the question and establish what it is that we are being asked to calculate or discover. Using these steps for each new question we attempt can, with practice, make solving physics problems extremely satisfying and rewarding.

- **Step 1: Keywords**
  Are there words in the question that contain additional information about the problem? Frictionless; light; uniform… Highlight these words so that they stand out — they will help to simplify the solution and allow us to neglect concepts that we don't need to consider.

- **Step 2: Draw a diagram**
  The action of drawing a diagram helps us to digest and summarise the information in a question. Drawing a diagram will save time and effort later and is the key to finding a solution. When drawing the diagram, label the quantities that are given with symbols (e.g. u for a velocity, d for a displacement); staying in symbols rather than using numbers is vital — see Step 4.

- **Step 3: Key concepts & mathematics**
  Fluency with mathematical rearrangements is essential but we need to

be sure that we are logical with our approach and consider the physical concepts that we need before throwing algebra at the problem. Identify which concepts are relevant to the problem. Write down the relevant physical principles and equations that might be useful — particularly if they connect the quantities that are given in the question.

- **Step 4: Stay in symbols**
  Even if the question gives numerical values, represent each of them with a symbol. This may appear to overcomplicate the problem, but it really helps when checking the solution or trying to find a mistake.

  For example, imagine that as part of the calculation we want to find the magnitude of the displacement, $|\underline{d}|$, of a cyclist who has travelled 13 km West and 5.0 km North.

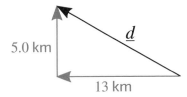

Figure 1: The displacement of a cyclist who has travelled 13 km West and 5.0 km North.

You identify that you need to use Pythagoras' theorem

$$|\underline{d}| = \sqrt{144} = 12 \, \text{km} \qquad\qquad \text{WRONG} \qquad\qquad (1)$$

The numbers hide information about the calculation that letters would not. Let $w = 13$ km and $n = 5$ km

$$|\underline{d}| = \sqrt{w^2 - n^2} = 12 \, \text{km} \qquad\qquad \text{WRONG} \qquad\qquad (2)$$

You can see straight away that the square of the numbers have accidentally been subtracted rather than added.

- **Step 5: Check dimensions, then put in numbers, check if reasonable**
  Before putting numbers into an algebraic expression, experienced physicists check whether their answer has the correct *dimensions* (you may have seen this before as checking your units). For example, if we are trying to find a quantity of time, *each term* in our expression must have dimensions of *only* time — no length, or mass or charge!

  Imagine that we are trying to calculate how fast the Earth travels around the Sun and we know that the radius of the Earth's orbit is $r = 1.50 \times 10^{11}$ m and that it takes $t = 365.25$ days. We make a mistake and write down that the distance travelled by the Earth is $2\pi r^2$. If we check our dimensions on both sides of the equation, we can see that this is incorrect.

$$\text{speed} \quad = \quad \frac{2\pi r^2}{t} \tag{3}$$

$$\text{Dimensions:} \qquad \frac{[\text{L}]}{[\text{T}]} \quad \neq \quad \frac{[\text{L}]^2}{[\text{T}]} \tag{4}$$

The length dimension, $[\text{L}]$, that we have on the top of the left hand side of expression (4) does not match the length squared, $[\text{L}]^2$ on the top of the right hand side of expression, so we determine that there is a mistake with our r^2 part of the expression.

Lastly, we put in the numbers and, vitally, check the magnitude of our answers – are they reasonable?

Now we have the steps, we apply them to two example problems from Levels 3 and 5 of isaacphysics.org.

Example Solution - Level 3 Dynamics: Pop-up Toy
isaacphysics.org/s/i80Q9P

> **Q:** A pop-up toy consists of a head and sucker of combined mass $m = 1.5$ kg stuck to the top of a light spring of natural length $l_0 = 0.30$ m and spring constant $k = 250$ N m^{-1}. The centre of mass of the system can be taken to be at the top of the spring. The spring is compressed to length $l_1 = 0.10$ m when the pop-up toy is stuck to the ground.
>
> What height above the ground does the bottom of the unstretched spring jump to when it is smoothly released?

Step 1: Keywords

Highlight the key words in the text above.
What are the key words in the question that will help you to draw your diagram and understand the physical concepts that might be useful? Note also what the question is asking for.

Solution:
- combined mass at the top — we can consider the sucker and mass as one object of mass m at the top of the spring.
- centre of mass — the point at which we should consider the weight of the combined mass to act.

- light — we can consider the spring to have no mass or weight, it is negligible.
- spring — is extensible.
- smoothly released — the toy is not caused to jolt in anyway so there is no energy converted to heat or sound; we conserve kinetic and potential energies.
- **Q:** Height of the bottom of the spring.

Step 2: Draw a diagram

Draw and annotate a diagram with all of the information from the question, in particular draw each stage of the toy's behaviour.

Solution:

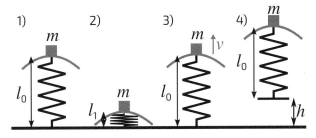

Figure 2: The 4 stages of the toy's behaviour. 1) Before compression. 2) Compressed and ready to jump. 3) Returned to its natural length, l_0, but now the mass and sucker has a velocity, v. 4) At maximum height, the velocity of the mass and sucker $= 0$.

Steps 3 & 4: Concepts, mathematics & symbols

Identify the concepts that you think might be useful in answering the question. Doing this gets you started. Some ideas might not get used.

Solution: Hooke's Law, spring constant, forces, work done, gravitational potential energy (GPE), elastic potential energy (EPE), kinetic energy (KE) and conservation of energy.

Method 1: Energy
How do you know whether to analyse a problem using forces (method 2) or energy? Using conservation of energy in this problem is simpler, as the total amount of energy remains the same throughout all the stages of motion and therefore we can consider just the beginning and end situations (stage 2 and stage 4).

Write down an expression for the conservation of energy. The total energy of the toy at stage 2 is equal to its total energy at stage 4.

- Take care to define a position of zero gravitational potential energy.

Solution: For the whole of this solution I will choose for the ground level to be the position of zero gravitational potential energy.

Total energy at stage 2 $=$ Total energy at stage 4

$$GPE_2 + EPE_2 = GPE_4$$

$$mgl_1 + \frac{1}{2}k(l_0 - l_1)^2 = mg(h + l_0)$$

$$h + l_0 = l_1 + \frac{1}{2mg}k(l_0 - l_1)^2$$

$$h = l_1 - l_0 + \frac{1}{2mg}k(l_0 - l_1)^2 \tag{5}$$

Method 2: Force and Work Done

Using a force method for the pop-up toy is more challenging because the force due to the spring is not constant throughout the motion. However, analysing the problem through forces should give us the same answer.

How does the force exerted by the spring on the mass cause the toy leap off the ground?

Solution: Hooke's Law tells us that the magnitude of the force, F, needed to compress a spring by an amount x, is given by $F = kx$. To compress the spring by an additional small amount δx we need to do work **on** the spring. The amount of work done is $\delta W = F\delta x$. As δx becomes very small (tends to zero), δW tends to dW.

We can then find the total work done to compress the spring from stage 1 to stage 2 by integrating with respect to dx between the limits of the compression at stage 1 to the compression at stage 2.

Stage 1, compression $= 0$

Stage 2, compression $= (l_1 - l_0)$.

The total work done on the spring between these two stages is then released between stages 2 and 4, and converted to gravitational potential energy.

Work done from 1 to 2 $=$ change in GPE from 2 to 4

$$\int_0^{(l_1 - l_0)} (kx)dx = GPE_2 - GPE_4$$

$$\left[\frac{1}{2}kx^2\right]_0^{(l_1 - l_0)} = mgl_1 - mg(h + l_0)$$

$$h + l_0 = l_1 + \frac{1}{2mg} k(l_1 - l_0)^2$$

$$h = l_1 - l_0 + \frac{1}{2mg} k(l_1 - l_0)^2$$

Our result is indeed consistent with that of method 1, equation (5).

This is not the only force method we could use, the problem can also be solved using Newton's Second Law to equate the resultant force on the toy, at a general time, to its acceleration. The acceleration must be written in terms of $\frac{dv}{dx}$ rather than $\frac{dv}{dt}$ so that we can then integrate to find an expression for v^2 which can be related to the height h.

Step 5: Dimensions & numbers

The question asks for a height, which has dimensions of length $[L]$ — each term on the right of our expression (5) should therefore also have dimensions of length.

Solution: Not all of the terms have dimensions that we can just write down; for example, what are the dimensions of k and g?

$$g = \text{acceleration} = [L][T]^{-2}$$

$$k = \frac{\text{force}}{\text{length}} = \frac{[M][L][T]^{-2}}{[L]} = [M][T]^{-2}$$

$$h = l_1 - l_0 + \frac{1}{2mg} k(l_0 - l_1)^2$$

$$= [L] + [L] + \frac{[1]}{[M][L][T]^{-2}} [M][T]^{-2}([L])^2$$

$$= [L] + [L] + [L] = \text{correct}$$

Now we substitute the values given in the question and consider carefully the number of significant figures we should give in our answer.

$m = 1.5$ kg, $l_0 = 0.30$ m, $l_1 = 0.10$ m, $k = 250$ N m^{-1} and $g = 9.81$ m s^{-2}.

Solution:

$$h = l_1 - l_0 + \frac{1}{2mg} k(l_0 - l_1)^2$$

$$h = 0.10 - 0.30 + \frac{1}{2 \times 1.5 \times 9.81} \times 250 \times (0.30 - 0.10)^2 = 0.14 \, \text{m}$$

Is this answer reasonable? 14cm is indeed a realistic height for the bottom of the spring to reach.

Example Solution - Level 5 Statics: Prism
isaacphysics.org/s/i80Q9P

> A prism has a cross section that is an isosceles triangle. It has a unique angle
> of $30°$, and a mass of $m = 100$ g. You wish to lift it by touching the upper
> two faces only.
>
> **Q:** If the coefficient of friction between the prism's surface and your skin
> is $\mu = 0.400$, what is the minimum normal force you need to apply to each
> face in order to support the prism?

Step 1: Keywords

Highlight the key words in the text above.
What are the key words in the question? Note exactly what the question asks.

Solution: isosceles, friction, minimum, normal, unique angle.
Q: We are asked for the minimum normal force applied to **each** face.

Step 2: Draw a diagram

Redraw and annotate a diagram of the prism.

- Label your angle θ to stay in symbols until the end of the calculation.
- Annotate the diagram with **all** the forces acting on the prism. Think carefully about the forces that act **on the prism** as opposed to those **on your fingers**.

Solution: To add the frictional force on the prism we think about the direction that the prism wants to fall — we know that friction will oppose this motion. Using Newton's third law, the frictional force on our fingers must be equal in magnitude but opposite in direction to that on the prism.

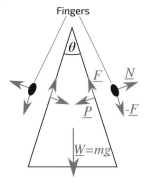

Figure 3: The forces on the isosceles prism and on your fingers as you lift the prism and hold it in static equilibrium. The forces on the right hand side have been labelled, those on the left will have the same magnitudes.

Steps 3 & 4: Concepts, mathematics & symbols

What concepts may we need to solve this problem?

Solution: Newton's first law, Newton's third law and the law of friction.

Write down the magnitude of the frictional force in terms of the coefficient of static friction μ and the normal component, P of the force that you apply to each upper face of the prism.

- Note that the force that you apply to the prism is equal and opposite to the reaction force of the prism on your finger, e.g. $N = -P$, by Newton's third law.

Solution: From the law of static friction: $F \leq \mu N = \mu P$

To find the minimum value of P we use the maximum value of F: $F = \mu P$

Apply Newton's First Law vertically.

Solution: From Newton's First Law and resolving forces vertically:

$$2F \cos (\theta/2) - (W + 2P \sin (\theta/2)) = 0$$

Using $F = \mu P$: $2\mu P \cos (\theta/2) - (W + 2P \sin (\theta/2)) = 0$

By rearranging the latter expression we can write down an expression for the force we apply, P, in terms of μ, the angle θ and the weight of the prism mg.

Solution:

Applied normal force, $P = \dfrac{W}{2\left[\mu \cos(\theta/2) - \sin(\theta/2)\right]}$

Step 5: Dimensions & numbers

We are asked to find a force so we should check that the dimensions of our expression are the same as that of a force, $F = [M][L][T]^{-2}$

Solution:

$$\frac{W}{2\left[\mu \cos(\theta/2) - \sin(\theta/2)\right]} = \frac{[M][L][T]^{-2}}{\text{dimensionless quantities}}$$

Our expression is indeed dimensionally correct.

Now using the values given and taking care with the number of significant figures in our answer, $m = 100$ g, $g = 9.81$ m s^{-2}, $\mu = 0.400$ and $\theta = 30.0°$.

Solution:

$$P = \frac{W}{2\left[\mu \cos(\theta/2) - \sin(\theta/2)\right]} = \frac{0.100 \times 9.81}{2\left[0.400 \cos(15.0) - \sin(15.0)\right]} = 3.85 \text{ N}$$

Consider whether this is a reasonable answer by comparing it with the weight of an everyday object.

Level 1: Mechanics

Each link by a topic heading will take you to nine more similar problems.

Statics
isaacphysics.org/s/Dbv5Dj

Q: Bed of Nails

A uniform rod of weight $500\,$N is supported by two pegs at either end of the rod.

Part A: Draw a free body diagram showing the forces acting on the rod. What is the magnitude of the forces exerted by each of the pegs on the rod?

Part B: If there are now eight pegs evenly spaced along the rod to support its weight, what force is applied by each peg on the rod?

Part C: Using the previous answers, explain how it is possible to lie on a bed of nails, but putting weight on one nail is extremely painful.

Dynamics
isaacphysics.org/s/CMjHTE

Q: A Hamburger

A hamburger has $2.2\,$MJ of chemical potential energy which can be released by eating it. (Take $g = 9.81\ \mathrm{m\,s^{-2}}$)

Part A: How much energy is needed to lift a $50\,$kg student upwards through a height of $0.40\,$m?

Part B: How many step-ups of $0.40\,$m would the student need to make in order to burn off the energy of a hamburger?

Part C: How many hamburgers would the student need to consume in order to reach the top of Mount Everest, a height of $8.8\,$km?

Part D: If the energy of a hamburger could be used by itself to propel itself upwards, how high could it rise? (The mass of a hamburger is $0.22\,$kg; assume this remains constant. Assume that g also remains constant).

Kinematics
isaacphysics.org/s/IUV78T

Q: A Strange Planet

A lost astronaut lands her spaceship on an unknown planet. She decides to work out the value of the acceleration due to gravity on this planet so that she can check her onboard computer and find out her location. She knows that on Earth (where $g = 9.81\ \mathrm{m\,s^{-2}}$ downwards) she takes $1.0\,$s to jump up and land again. On this planet, a jump takes $1.4\,$s.

What is the magnitude of the downward acceleration due to gravity on the strange planet? Assume she can jump up at the same speed on any planet.

Level 2: Mechanics

Statics

isaacphysics.org/s/B16Mzd

Q: Force on Table Legs

A uniform table consists of a circular wooden board of mass $m = 3.00$ kg resting on top of three vertical legs, each of mass $M = 0.500$ kg. The legs are equidistant from the centre of the table and form an equilateral triangle. (Take $g = 9.81$ m s^{-2}).

Part A: What is the magnitude of the reaction force from one of the legs on the tabletop?

Part B: What is the magnitude of the reaction force from the ground on one of the legs?

Dynamics

isaacphysics.org/s/4JKSmy

Q: The Lift

A lift, of mass 500 kg, is travelling downwards at a speed 5.0 m s^{-1}. It is brought to rest by a constant acceleration over a distance 6.0 m.

Part A: What is the tension, T, in the lift cable when the lift is stopping?

Part B: What is the work done by the tension whilst stopping the lift?

A lift, of mass 500 kg, is travelling upwards at a speed 5.0 m s^{-1}. It is brought to rest by a constant acceleration over a distance 6.0 m.

Part C: What is the tension, T, in the lift cable when the lift is stopping?

Part D: What is the work done by the tension whilst stopping the lift?

Kinematics

isaacphysics.org/s/L1TER3

Q: A Stoplight

A car is travelling along a road at constant speed $u = 7.00$ m s^{-1}. The driver sees a stoplight ahead change to amber and applies the brakes. Consequently, the car experiences a total resistive force equal to $\frac{1}{6}$ of its weight and stops just at the stoplight. Taking $g = 9.81$ m s^{-2}, how far does the car travel during the braking period?

Level 3: Mechanics

Statics

isaacphysics.org/s/T2B6VD

Q: Shelf and Brackets

A shelf of uniform density is supported by two brackets at a distance of $\frac{1}{8}$ and $\frac{1}{4}$ of the total length, L, from each end respectively.

Part A: Find the ratio of the reaction forces from the brackets on the shelf. Express your answer in the form of a decimal less than or equal to 1, to three significant figures.

Part B: The rectangular shelf is now replaced with a uniform right-angled triangular shelf (figure given on-line). The centre of mass of a triangle lies at a point $\frac{1}{3}$ of the perpendicular distance from the base to the tip.
Find the new ratio of the forces on the two brackets. Express your answer in the form of a decimal less than 1, to three significant figures.

Dynamics

isaacphysics.org/s/klHQYn

Q: A Moment of Rest

Two bodies, P and Q, of equal mass are travelling towards one another on a level frictionless track, with speeds u and u respectively, where $u > v$. They make an elastic collision. At some instant during the collision, P is brought instantaneously to rest.

What is the speed, v_Q, of Q at this instant?

Kinematics

isaacphysics.org/s/ktiMKD

Q: The Bolt Thrower

A castle wall has bolt throwers which fire a bolt horizontally at a speed v. In order to fire over the enemy's shields, the bolt must make an angle of at least $\theta = 45°$ to the horizontal when it hits the ground. The bolt throwers can be mounted at different heights in the wall and set to fire at different speeds.

Part A: Find the maximum range of a bolt fired from a height $h = 10$ m.

Part B: Find the speed required to reach the maximum range calculated in Part A.

Level 4: Mechanics

Statics
isaacphysics.org/s/YUWgEx
Q: Spring Triangle

A rod AB of length $d = 2.00$ m is fixed horizontally. Two light identical springs of spring constant $k = 14.0$ N m^{-1} are attached to the rod, one at each end. The loose ends of the springs are attached to each other at a point C and in this framework the springs are just taut. It is found that the angle made by one of the springs to the vertical $\alpha = 45.0°$. A metal ball is then suspended from the springs at C and the angle made by one of the springs to the vertical is found to be $\beta = 30.0°$.

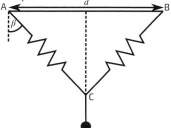

Figure 4: A metal ball suspended from two strings.

Taking the acceleration due to gravity as $g = 9.81$ m s^{-2}, what is the mass m of the ball?
(Multiple choice question – see options, concepts and feedback on-line.)

Dynamics
isaacphysics.org/s/4ox4q1
Q: A Ballistic Pendulum

A block of wood with a mass of $M = 2.5$ kg is suspended from fixed pegs by vertical strings $l = 3.0$ m long, in a set up known as a ballistic pendulum. A bullet with a mass of $m = 10$ g and moving horizontally with a velocity $u = 300$ m s^{-1} enters and remains in the block.

Find the maximum angle θ to the vertical through which the block swings.

Kinematics   isaacphysics.org/s/8BB7Rw
Q: Broken Cannon

A cannon on horizontal ground, at point C, is used to target a point T, 25 m behind a narrow wall. Unfortunately the cannon is damaged and can only fire at a
45° angle and at one speed. So, the only way to aim the cannon is by moving it
towards and away from the target. The gunners aren't sure if they can actually
hit the target.

Part A: If the cannonball leaves the cannon at $u = 35$ m s^{-1}; at what distance, d,
must the cannon be placed in front of the wall in order to hit the target, if the
wall is ignored and the target is at the same height as the cannon?

Part B: The wall is 15 m high. Does the cannonball actually go over the wall and
hit the target? If so, by how much?

SHM   isaacphysics.org/s/DZYEBC
Q: Car Suspension over Bumps

The suspension of a car may be considered to be an ideal spring under compression. When the driver, of mass m_1, steps into the car, of mass m_2, the vertical
height of the car above the road decreases by x. If the car is driven over a series
of equally spaced bumps, the amplitude of vibration becomes much larger at
one particular speed.

Explain why this occurs, and find an expression for the separation d of the bumps
if it occurs at a speed of v.

Angular Motion
isaacphysics.org/s/mcIYvI
Q: Hammer Throwing

At a hammer throwing event, the speed that an athlete is spinning is such that they complete one revolution in 0.50 s. Most of the weight of the hammer is in the ball, so it can be approximated as a point mass $m = 7.0$ kg at a distance $l = 1.2$ m away from the end of the handle.

The athlete stretches their arms out such that they are holding the end of the handle of the hammer a distance $d = 0.60$ m away from their axis of rotation. Assume that the hammer has no vertical motion.

Part A: What is the linear speed of the ball at the end of the hammer?

Part B: What is the kinetic energy of the hammer?

Circular Motion
isaacphysics.org/s/GJMwfO
Q: Car on a Roundabout

A car approaches a level roundabout of radius $R = 10$ m.

Part A: What is the maximum speed that the can travel around the roundabout without slipping, if the coefficient of friction between the road and car is $\mu = 0.80$?

A racing car approaches a corner of radius $R = 50$ m, banked at an angle $\theta = 30°$ to the horizontal.

Part B: What is the maximum speed at which the car can travel if it is icy so there is no friction between the wheels and the road?

Part C: What is the maximum speed at which the car can travel around the banked corner and not slip if the road is rough, with coefficient of friction $\mu = 0.55$?

Part D: What is the minimum speed at which the car needs to travel around the banked corner when the road is rough, with coefficient of friction $\mu = 0.55$?

Level 4: Fields

Electric Fields

isaacphysics.org/s/q6IUoh

Q: Field between Parallel Plates

Two large, parallel metal plates are placed 0.130 m apart. The potential difference between the plates is 150.0 V .

Part A: Electric field strength Calculate the value of the electric field strength between the plates.

Part B: Force on electron Calculate the size of the force on an electron, of charge -1.60×10^{-19} C, placed half way between the plates.

Part C: Force variation How does the force on the electron vary as it moves from the negative plate to the positive plate?

Part D: Potential variation What is the magnitude of the potential difference between two points that are positioned between the parallel plates a distance of $\frac{1}{4}$ and $\frac{3}{4}$ of the way in from one of the plates? Give your answer to 3 significant figures.

Magnetic Fields

isaacphysics.org/s/RjHZfp

Q: Electron in a Magnetic Field

An electron, of charge -1.6×10^{-19} C and velocity 2.35×10^7 m s^{-1}, moves perpendicularly to a magnetic field with magnetic flux density 3.4×10^{-2} T.

Calculate the size of the force on the electron.

Gravitational Fields

isaacphysics.org/s/b9I6Cv

Q: Escaping from the Moon

The mass of the Moon is 7.4×10^{22} kg and its radius is 1.7×10^6 m. Newton's gravitational constant is $G = 6.67 \times 10^{11}$ N m^2 kg^{-2}. Assume a spherical Moon.

Part A: Calculate the gravitational field strength at the Moon's surface.

The mass of the Earth is 6.0×10^{24} kg and its radius is 6.4×10^6 m. Assume that the Earth is spherical.

Part B: What is the escape velocity on the Moon as a fraction of the Earth's escape velocity?

Gravitational Fields
isaacphysics.org/s/b9I6Cv
Q: Two Orbiting Masses

Two masses of mass m and M are placed with their centres a distance r apart.

Part A: Attractive Force Between Masses Find an expression for the magnitude of the gravitational attractive force between the two masses.

Part B: Centre of Mass In terms of r, how far away from the larger mass M is the centre of mass of the two-mass system?

Part C: Circular Orbit Consider both masses to be performing circular orbits about the centre of mass of the system. Find the speed of the smaller mass in terms of m, M, r and G.

Combined Fields
isaacphysics.org/s/kGlyBR
Q: Where do the Charges Go?

An region of space has an electric field acting to the right and a magnetic field acting "into the screen" as shown in fig. 5.

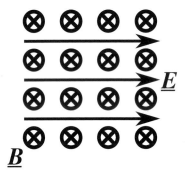

Figure 5: Electric and magnetic fields.

If a single positively charged particle were placed at rest into these fields, in which direction would it move after a short finite time? What direction would a separate negatively charged particle move after a short finite time if it alone were placed at rest into these fields?
(Multiple choice question – see on-line for choices, concepts, hints, video.)

Level 5: Mechanics

Statics

isaacphysics.org/s/ILSBvf

Q: Festival Banner

A festival banner of mass m is to be hung between two buildings of equal height, a distance l apart, by a light inextensible wire of length λl, where λ is a constant. So that it is not stretched, the banner must be placed such that there is equal tension on each length of wire supporting it.

The wire is attached to the roof of the building on the left, and to the side of the building on the right at a height h above the ground. The banner needs to be hung at a distance y vertically below the level of the buildings' roofs, and at a horizontal distance x from the building on the left.

Part A: Length of wire By considering the banner as a point mass, find the length of the wire, λl needed, as a function of l, x, y and h.

Part B: Vertical distance to banner Find y in terms of x and λ.

Part C: Tension in the wire If $\lambda = 2$, find the tension T in the wire.

Dynamics

isaacphysics.org/s/NhtQS1

Q: Picking up a Chain

Part A: Tension in the chain A chain of mass $\lambda = 2.0 \, \text{kg m}^{-1}$ per unit length has a tension $T(0)$ applied to one end. First consider the case where the chain is hanging vertically and is of total length $L = 10$ m.
Give an expression for the tension and calculate the tension, $T(5)$, at $x = 5$ m, where $x = 0$ is at the upper end of the chain.

Part B: Raising the chain A long length of chain is now placed on a smooth flat surface, and one end is raised at a constant speed $v = 5.0 \, \text{m s}^{-1}$ by a tension $T(0)$.
In the instant that the last piece of chain is raised, i.e. when length, $L = 10$ m is suspended, what is $T(0)$?

Part C: Calculating power What is the power input by $T(0)$ at this point?

Part D: Change in energy What is the rate of change of energy stored in the system as kinetic energy or gravitational potential energy at this point?

Kinematics   isaacphysics.org/s/XarSrv

Q: Crossing a River

A river has banks at $x = 0$ and $x = a = 20$ m, and flows parallel to its banks with a speed $u = kx(a - x)$, where $k = 0.010$ m^{-1} s^{-1}.

The captain of a ship wishes to moor as far upstream on the opposite bank as possible. If his initial speed is $v = 0.50$ m s^{-1}, what angle θ to the x direction should he set off at? You should assume that the ship's power output is constant, and that it always thrusts in the same direction relative to the banks.

SHM   isaacphysics.org/s/da09Es

Q: Accuracy of an SHM Approximation

A simple pendulum of length L is released from rest from an angle of θ_0 to the vertical. The mass on the end is assumed to be a point mass.

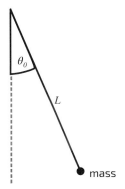

mass   **Figure 6:** A simple pendulum.

Part A: SHM approximation. Assuming that the pendulum undergoes simple harmonic motion, find an expression for its maximum speed.

Part B: Exact speed. Using conservation of energy, find an exact expression for this speed.

Part C: Compare results. Find the percentage error in the SHM approximation of the maximum speed compared to the true value when $\theta_0 = 0.20$ rad.

Angular Motion   `isaacphysics.org/s/HyTNE6`

Q: Destroying the Moon

Orbital bombardment is the name for the process where asteroids could be nudged out of their current stable orbit and sent careering into a target. One day a particularly eccentric spaceship captain decides that he doesn't like the Earth's Moon and attempts to destroy it by sending an asteroid to collide with it.

He chooses an asteroid with exactly the same mass as the Moon and alters its orbit so that it will hit the Moon. Only after the asteroid is launched does one of his staff point out that Earth might be in danger as well.

The asteroid collides with the Moon head on, and before the collision it is moving slower than the Moon. The resulting cloud of debris stays together. Ignore the gravitational attraction between the Moon and the asteroid.

Assuming the Moon to be initially in a circular orbit at radius $r = 380\,000$ km from the centre of the Earth, using the acceleration of free fall at the Earth's surface $g = 9.81\,\mathrm{m\,s^{-2}}$, and given that Earth's radius is $R = 6378$ km, find the maximum speed the asteroid can hit the Moon without the cloud colliding with the Earth.

Circular Motion   `isaacphysics.org/s/dS3Pg1`

Q: A Particle in a Bowl

A smooth hemispherical bowl of radius $a = 15$ cm is placed with its axis of symmetry vertical, and a particle of mass $m = 50$ g moves in a horizontal circular path on the inside of the bowl with a speed v. The plane of this circle is situated half way down the axis.

Part A: Find the value of v.

Part B: Find an expression for the reaction, N, of the bowl on the particle. What is the magnitude of N?

Level 5: Fields

Electric Fields

isaacphysics.org/s/G7rS6I

Q: Inkjet Printing

In some methods of inkjet printing, different droplets are given different amounts of charge as they are fired horizontally out of the nozzle. These droplets are then deflected by different amounts by a deflection voltage that is applied across two electrodes.

In one printer the deflection voltage is $480\,\text{V}$, the length of the deflecting electrodes is $x = 1.0$ mm and the distance between them is $d = 0.20$ mm. The electric field between these electrodes can be assumed to be constant. The droplets produced are approximately spherical with a density of $\rho = 1050$ kg m^{-3} and a radius of $a = 30$ μm. They leave the nozzle with an initial speed of $u = 5.5$ m s^{-1}.

What is the difference in the magnitude of charge between two droplets that hit the paper at the top and bottom of a letter 'I' of height $h = 2.0$ mm a distance of $L = 0.50$ cm away from the edge of the deflecting electrodes? Gravitational effects can be ignored.

Magnetic Fields

isaacphysics.org/s/Nb5XzE

Q: Magnetic Inclination

The Earth's magnetic field is generally not parallel to its surface. Near Cambridge, the magnetic field points downwards at an angle of $66°$ to the ground. Three particles of equal charges move with the same speed in three different directions: directly towards magnetic North, at right angles to this along the ground (East), and vertically. The Lorentz forces on these particles are labelled as F_{North}, F_{East} and F_{Vertical} respectively.

Order these forces in decreasing magnitude.
(Multiple choice question. See choices and feedback on-line.)

Gravitational Fields

isaacphysics.org/s/y0Df63

Q: Estimating Gravity

The value for the acceleration due to gravity, $g = 9.81$ m s^{-2}, is only valid near the Earth's surface. Further away from the Earth the value of g varies with the distance from the Earth's surface h.

Part A: Gravity strength at edge of atmosphere. The Earth's atmosphere has a thickness of about 700 km. By finding a power series of g in h, estimate the ratio of the acceleration due to gravity at the top of the Earth's atmosphere to that at its surface, given that the radius of the Earth is approximately $R = 6378$ km. You should assume that $h \ll R$ such that $(h/R)^2 \approx 0$. Give your answer as $\frac{g_{atmos}}{g_{surface}}$ to 3 significant figures.

Part B: Estimate error. Estimate the error in your previous answer. Give your answer to 3 significant figures.

Combined Fields   isaacphysics.org/s/rRP40E

Q: Fields in a Triangle

The diagram below shows a triangle ABC in which CA = CB = r. In the first case, an electric field is created by placing point charges $-Q$ and $+Q$ at A and B respectively. In the second, two equal point masses m are situated at A and B to create a gravitational field in the region.

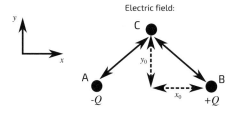

Figure 7: Diagram showing the triangle with electric and gravitational fields.

Part A: Direction of electric field. In which direction, relative to the x- and y-directions, does the resultant electric field at C act?
(Multiple choice.)

Part B: Direction of gravitational field. In which direction, relative to the x- and y- directions, does the resultant gravitational field at C act?
(Multiple choice.)

Part C: Magnitude of electric field. If $Q = 1.60 \times 10^{-19}$ C, $x_0 = 2.0$ mm and $y_0 = 3.5$ mm, what is the magnitude of the electric field at C?

Part D: Magnitude of gravitational field. If $m = 1.0$ kg, $x_0 = 2.0$ mm and $y_0 = 3.5$ mm, what is the magnitude of the gravitational field strength at C?

Level 6: Mechanics

Statics

isaacphysics.org/s/i2mLQr

Q: Hanging a Non-uniform Bar

A non-uniform bar of mass m is hung horizontally between two walls, using two light ropes attached to the ends. One of the ropes makes an angle $\theta = 36.9°$ to the vertical, and the other makes an angle $\phi = 53.1°$ to the vertical.

If the bar is $l = 1.00$ m long, how far away from the closest edge is the centre of mass?

Dynamics

isaacphysics.org/s/aA2xdL

Q: Space Justice

A police spaceship, of mass $m = 10000$ kg, travelling at a speed $u = 2.00$ km s^{-1} needs to arrest another ship travelling ahead of it at 2.50 km s^{-1}. The police spaceship is capable of splitting itself into two equal parts and supplying them with kinetic energy from a single reserve of 2.00 GJ.

Part A: Calculate the speed of the part of the police spaceship that remains pursuing the other ship. Was the energy reserve sufficient?

Part B: If the reserve is sufficient; how long would it take for the police spaceship to catch the other craft, given that it was initially 3000 km behind?

Kinematics

isaacphysics.org/s/WgAvf5

Q: The Bouncing Ball

A ball is dropped from rest at a height h_0 and bounces from a surface such that the height of the nth bounce, h_n, is given by $h_n = \alpha h_{n-1}$, where h_{n-1} is the height of the previous, $(n-1)^{\text{th}}$ bounce. The factor α has value $0 \leq \alpha \leq 1$.

Part A: How far does the ball travel before coming to rest, given that $\alpha = 0.25$ and $h_0 = 3.0$ m?

Part B: How long does the ball take to cover this distance?

SHM   isaacphysics.org/s/yV9R96
Q: Gravtube

Imagine a tube had been drilled straight through the centre of a uniform spherical planet. The planet has a radius R and the acceleration due to gravity at its surface is g. An object of mass m is released from rest at one end of the tube. From Gauss's theorem, the gravitational force on an object of mass m inside a uniform massive spherical body (in this case, a planet), is given by $F = -\dfrac{GMm}{r^2}$ where r is the distance of the small mass from the centre of the planet and M is the mass of the planet that exists inside a sphere of radius r (ie. all the mass of the planet that is closer to the planet's centre than the mass m). The force is negative as it acts inwards, towards the point $r = 0$.

Show that the acceleration of an object inside this tube is of the form $a = -\omega^2 r$ and so the object moves with simple harmonic motion.

Part A: What is the time period of the resulting oscillation if $g = 6.00$ m s^{-2}, and $R = 1200$ km?

A satellite is placed in a circular orbit around the same planet, so that it is orbiting just above the ground (ie, at a radius R). The centripetal acceleration of an object of mass m, in a circular orbit at a radius r, is given by $a_c = \omega^2 r$, where ω is the angular velocity of the mass in orbit.

Part B: Find an expression for ω and calculate the time period of the orbit at this radius?

Angular Motion   isaacphysics.org/s/4k2no4
Q: Rising Hoop

Two beads, each of mass m, are positioned at the top of a frictionless hoop of mass M and radius R, which stands vertically on the ground. The beads are released and slide down opposite sides of the hoop.

What is the smallest value of $\frac{m}{M}$ for which the hoop will rise up off the ground at some time during the motion?